An Integrated Assessment of China's Ecological Restoration Programs

Soji.

pleased to work w/ you.

An Integrated Assessment of China's Ecological Restoration Programs

Edited by

Runsheng Yin
Michigan State University, East Lansing, MI, USA

Editor
Runsheng Yin
Michigan State University
Department of Forestry
126 Natural Resources
East Lansing, MI 48824-1222
USA
yinr@msu.edu

ISBN 978-90-481-2654-5   e-ISBN 978-90-481-2655-2
DOI 10.1007/978-90-481-2655-2
Springer Dordrecht Heidelberg London New York

Library of Congress Control Number: 2009926830

Cover illustration: Cover photo taken by Mr. Zhendong Feng of Wuqi Office of Land Retirement and Conversion. It shows a watersheld in the Loess Plateau region, after converting crop fields on slopes to grass and tree coverage.

Printed on acid-free paper

Springer is part of Springer Science+Business Media (www.springer.com)

Foreword

When I visited China in 2006 on a US mission that examined forestry and the forest sector I was struck with the obvious improvements in the forest situation compared to my prior visit some 16 years earlier. Areas of forest restoration were visited and I was impressed with the data that suggested China's forest had made a remarkable recovery and indeed substantial expansion since the 1970s. This volume suggests not only that my anecdotal impressions were reasonably accurate for forest but that substantial progress is being made across a number of ecological areas.

The book was developed out of papers presented to a 2007 international symposium in Beijing that discussed aspects of several ecological restoration programs initiated by the Chinese government beginning in the late 1990s. The volume, edited by Runsheng Yin, covers a wide range of issues in its 14 chapters. The studies highlight recent research advances in assessing China's ecological restoration programs. Although each of these chapters is a stand-alone piece, the volume progresses in an orderly manner from an articulation of the challenges, a review of the literature, the presentation of research methodology, the modeling of the driving forces of some of the changes, empirical estimates of some of the changes associated with the various activities, and finally an assessment of the effects of these ecological restoration programs on the socio-economic aspects of a location or site.

The specific programs examined in the volume include those focused on forest protection, slopping cropland retirement, desertification control wildlife conservation, shelter belt development, and soil erosion avoidance. These programs have resulted in the rehabilitation of forest and grasslands, the restoration of farmlands, and a host of other ecological improvements. In some cases the land cover change, data and experience from the 1970s into the first decade of the 21st Century are examined. In many cases the results are quite positive. Monitoring has revealed, as noted, the dramatic increase in forest cover throughout much of China. Also, in the post 2000 period China has achieved the stabilization of farmlands and wet lands in certain regions, reversing the earlier declining trends. The studies also assessed the effects of the programs on soils, sequestrated carbon, land use and water in various regions of China.

The volume is extraordinary for the breath of its coverage. Not only are natural systems assessed, but so are social and economic systems. Data are collected and evaluated, for example, on economic condition, household income, and employment. The role of off-farm employment, for example, is shown to vary considerably but often can be quite important in the agricultural sector. The methodological approaches covered in the volume include information obtained from mapping, remote sensing, a host of on-the-ground procedures, as well as various socio-economic data. The analytical approaches used in the chapters are even more impressive covering theoretical and mathematical approaches and approaches that depend on household sampling data. The studies include an assessment of productivity changes in the crop sector, which shows that over time the output trend has increased despite a reduction in the area of croplands.

The studies provide a substantial improvement over the common practice of simply reporting on projects undertaken and accepting uncritically the assessment of the government. This volume assesses the degree of success of the various programs using empirical data that show some restoration programs have made significantly positive impacts over the relatively short time periods. Importantly, however, the authors are not naïve and counsel for care and vigilance in the development and administration of programs for the future.

This volume fills an important gap in our knowledge of the effects of ecological restoration programs upon a variety of Chinese lands and ecosystems. Sources outside of China have tended to stress the ecological degradation that has occurred. These studies indicate that there are serious efforts and some successes in reversing that trend. The superficial impressions of improvements and restoration that one might have based on selected observations are buttressed with data, facts, and analysis, suggesting that the improvements are, indeed, real. The volume's coverage is not confined to one sector. Restoration is being accomplished in a host of sectors. Finally, not only is this volume useful in conveying the general sense of ecological improvements, but it provides examples of useful analytical approaches and innovative research advances that assist in assessing the effectiveness of various programs and policies in China and elsewhere.

Washington, DC   Roger A. Sedjo

Preface

Since the late 1990s, the Chinese government has launched several unprecedented ecological restoration campaigns to deal with the increasingly severe problems of soil erosion, flooding, dust storms, and habitat loss, among others. They include the Sloping Land Conversion Program (SLCP), the Natural Forest Protection Program (NFPP), and the Desertification Combating Program around Beijing and Tianjing (DCBT). While there have been studies of these programs, many questions remain to be examined concerning why the programs have been initiated, whether they have been effectively implemented, what their induced socioeconomic and ecological impacts are, and how their performance can be improved. To address these overarching questions, my colleagues from the Forest Economics and Development Research Center (FEDRC) of the Chinese State Forestry Administration and Michigan State University (MSU) and I organized an international symposium in Beijing, on October 19, 2007. It included invited speeches and panel discussions as well as refereed presentations, many of which featured ongoing research projects undertaken by scholars from MSU, FEDRC, and their partners.

After the symposium, I decided to edit a book based on the conference presentations, and the presenters and their associates have worked hard to prepare their manuscripts for the book in a timely manner. Our consensus was that getting these papers quickly published in a single volume would highlight the research advances in assessing China's ecological restoration programs and thus provide a major source of literature, conceptual frameworks, models, and tools for continued assessment of the programs, particularly integrated assessment of the socioeconomic and environmental impacts. If time proves that this book has indeed made a valuable contribution in this regard, then our intention will have realized, for which my colleagues and I will be grateful. As its editor, I will take responsibility for any not so well articulated or elaborated arguments and language errors.

The book, containing 14 chapters, addresses a wide range of issues. It begins with an up-to-date description of China's ecological restoration programs (ERPs, Chapter 1) and an extensive survey of the literature that has assessed the implementation efficacy and impact significance of these programs (Chapter 2). Then it presents the research methodology of integrative assessment that my collaborators and I have adopted (Chapter 3) and a case study of detecting the land use and land cover changes (LUCC) induced by implementing the programs (Chapter 4).

Chapters 5 and 6 are devoted to modeling the driving forces of the historical LUCC in the upper Yangtze. Chapters 7 and 8 are attempts to quantify the potential ecological impacts on carbon dynamics and soil erosion. The remaining chapters are efforts to estimate the socioeconomic impacts of the programs, as reflected in income, employment, and other dimensions of livelihoods.

In describing China's ERPs, Chapter 1 covers their initiation, implementation, and challenges. Overall, it appears that with the substantial government investment, tremendous progress has been made in implementing the ERPs. For instance, the forest area and volume have expanded significantly with the implementation of the NFPP. Similarly, under the SLCP a large amount of degraded farmland and grassland has been rehabilitated, and the forest and grass coverage has expanded substantially. As a result, the ecological and socioeconomic conditions have seen improvement. To complete their implementation successfully and to fundamentally improve the ecosystem functions and services, however, the authors argue that it is essential for China to embrace a more balanced and comprehensive approach to ecological restoration, adopt better planning and management practices, emphasize local people's active engagement, strengthen the governance of project implementation, establish an independent and competent monitoring network, and conduct timely and high-quality assessments of the program effectiveness and impacts.

In addition to reviewing the literature on assessing the implementation efficacy and impact significance of the ERPs, Chapter 2 also outlines several directions for future research. In a nutshell, it finds that: (1) the implementation effectiveness has not been examined as extensively as the impact significance; (2) efforts to assess the impact significance have concentrated on the SLCP, particularly its socioeconomic effects: growth of income, alternative industry, and employment, and likelihood of re-conversion; and (3) most of the socioeconomic studies are based on rural household surveys, and discrete choice and difference in differences models. Future work should thus pay more attention to the NFPP and other programs, and the environmental impacts and the implementation effectiveness of all of them. To these ends, the authors recommend that analysts gather more field data regarding the evolving ecosystem conditions and socioeconomic information of higher aggregation, and conduct their research across scales, with better application of geospatial technology and more effective collaboration.

In presenting their research methodology of integrative assessment, the authors of Chapter 3 first call for embracing both environmental and socioeconomic changes and engaging investigations at multiple scales and through interdisciplinary collaboration with expertise from ecology, economics, hydrology, and geospatial, climate, and land change sciences. Echoing some of the recommendations made in Chapter 2, they further argue that the deployment of geospatial capability, the use of longitudinal data, and the connection between science and policy should be the hallmarks of an integrative assessment. Then, they outline their general approach and specific models to quantifying the environmental and socioeconomic impacts of the ERPs, and to addressing the issue of how to overcome the challenges in generating the data needed for executing various empirical tasks. The authors hold that the adoption and application of this methodology will lead to a more rigorous and systematic assessment as well as implementation of the ecological restoration programs in China.

Chapter 4 reports the land cover changes in northeast China from the late 1970s to 2004. Even though land use statistics from governmental sources exist, their availability and reliability are not promising. Coupled with the fact that the geospatial technology is now not only well advanced but also cost effective, this led to the authors' decision to identify the regional LUCC using remote sensing and the Geographic Information System. Their results are very encouraging. It is found that while forestland and wetland were greatly reduced until 2000 due to farmland expansion and urbanization, spurred by the population and economic growth, their declining trends have been revered most recently. Meanwhile, built-up land has continued to increase. Further, the land cover changes occurred primarily in areas with low elevation and gentle slope. These findings suggest that the forest and wetland protection and restoration projects have indeed taken effect. However, the authors warn that there remains a long way to go before the ecosystems are greatly recovered and can function in the way that society expects.

Chapters 5 and 6 are attempts to modeling the driving forces of the LUCC in the upper Yangtze basin, aiming to shed light on human impacts on the LUCC and environmental conditions. In Chapter 5, the authors use a fractional logit model to determine the effects of social, economic, and institutional factors on the changes of cropland, forestland, and grassland. Based on a panel dataset covering 31 counties over four time periods from 1975 to 2000, they show that population expansion, food self-sufficiency, and better market access drove cropland expansion, while industrial development contributed significantly to the increase of forestland and the decrease of other land uses. Similarly, stable tenure had a positive effect on forest protection. In addition to highlighting the main LUCC drivers, this chapter also illustrates the limitations of the conventional choice models. To gain a better knowledge of the complex interactions of human and natural drivers underlying the LUCC, Chapter 6 creates a system of simultaneous equations to capture the dynamic linkages, feedback, and endogeneity of the determinants. This novel structural model of land use focuses on the multiple dimensions of agriculture – not only cropland use itself, but also grain production, soil erosion, and related technical change. The results show that technical change plays an important role in supplying food on a limited cropland; limiting cropland expansion in turn reduces soil erosion, which then benefits grain production in the longer term. It is also found that policies and institutions have significant impacts on land use change and the status of soil erosion. Together, these results carry some important implications for sustainable land use and ecosystem management. The alternative modeling strategies are beneficial to improved knowledge of the drivers.

Chapters 7 and 8 are estimations of the changes in carbon storage and soil erosion based on ecosystem simulation models. Quantifying the spatial and temporal dynamics of carbon stocks in terrestrial ecosystems and carbon fluxes between the terrestrial biosphere and the atmosphere is critical to understanding the regional patterns of carbon storage and loss. In Chapter 7, the authors use the General

Ensemble Biogeochemical Modeling System to simulate the terrestrial ecosystem carbon dynamics in the Jinsha watershed of the upper Yangtze basin, based on a unique combination of spatial and temporal dynamics of such major factors as climate, soil properties, nitrogen deposition, and LUCC. It shows that the Jinsha watershed ecosystems acted as a carbon sink during the period of 1975–2000, with an average rate of 0.36 Mg/ha/yr, primarily resulting from regional climate variation and local LUCC. Vegetation biomass accumulation accounted for 90.6% of the sink. While soil organic carbon loss before 1992 led to a lower net gain of carbon in the watershed, soils became a small sink thereafter. The carbon sink/source pattern had a high degree of spatial heterogeneity, with sinks associated with forest areas without disturbances and sources caused by stand-replacing disturbances. This underlies the importance of land-use history in determining the regional carbon sink/source pattern.

Land degradation caused by serious soil erosion has made the Loess Plateau one of the poorest regions in China. To improve the environmental conditions, the government has taken a number of measures there, including the SLCP. A natural question to ask thus is whether and to what extent it has actually accomplished the designed objective. Chapter 8 answers this question, concentrating on the soil erosion dynamics in the Zuli River basin. The authors do so by adopting a distributed soil erosion model to simulate the changes of water runoff and soil erosion induced by implementing the SLCP, with the assistance of remote sensing and GIS technologies for parameterization of the land surface attributes. Their simulations show that the improved ground cover, especially forestland and grassland, has resulted in an erosion reduction of 38.8%, compared to the mean level of the 1990s. On the other hand, the changed rainfall pattern has caused soil erosion to increase by 13.1%. In combination, the authors obtain a net decrease of soil erosion by 25.7% in recent years. This evidence suggests that China's ecological restoration efforts have effectively mitigated the regional water and soil loss.

Chapters 9–14 constitute studies of the different socioeconomic impacts of the ERPs – with different sample sites, datasets, and modeling approaches. While most of these chapters feature various versions of the difference-in-differences (DID) model and panel data, one chapter estimates the aggregate NFPP impacts using an input-output model and another measures the agricultural productivity change induced by the SLCP. The authors of Chapter 9 note that the ERPs are often designed with dual goals: to enhance ecosystem services and to alleviate poverty; however, reaching both can be challenging. If the household's supply of ecosystem services is not positively correlated with poverty, tradeoffs may exist between meeting the two goals. Moreover, even if the supply of ecosystem services and poverty were positively correlated, cost-conscious program managers need to adopt targeting approaches such that the poorer households with land that is less costly to set aside and provides a higher environmental benefit are selected. To explore strategies by which both the environmental and poverty alleviation objectives of a program can be achieved cost effectively, they develop a conceptual framework to understand the implications of alternative targeting when policy makers have both environmental and poverty alleviation goals. Using the SLCP as a study case, they then evaluate what factors determine the selection of program areas. Having demonstrated the heterogeneity of parcels and households, they examine the correlations across households and their parcels in terms of their potential environmental benefits, opportunity costs of participating, and the asset levels of households as an indicator of poverty. Finally, they compare alternative targeting criteria and simulate their performances in terms of cost effectiveness in meeting both the environmental and poverty alleviation goals when given a fixed budget. It is found that there is a substantial gain in the cost effectiveness of the program by targeting parcels based on the "gold standard," – targeting parcels with low opportunity cost and high environmental benefit managed by poorer households.

In Chapter 10, the authors hypothesize that in addition to participation status and household characteristics, the SLCP's impacts on income growth and labor transfer are determined by the local economic conditions, program extent, and political leadership; and the income impacts may vary from sector to sector. To test these propositions, they compiled a dataset of 600 households in three counties of the Loess Plateau region, with observations for times both prior to and after the program initiation (1999 and 2006), both aggregate and categorical incomes, and both participating and non-participating households. Using a simple DID model, they find that participation status, economic condition, program extent, and political leadership have all had significant impacts on household income and off-farm employment. Also, the effects of participation on crop production income, animal husbandry income, and off-farm income vary substantially. Obviously, these results have great policy meanings regarding how to improve the effectiveness and impacts of the ERP in China.

In Chapter 11, the authors assess the impact of the NFPP on local household livelihoods. To that end, they apply a series of policy evaluation microeconometric techniques to quantify the program's effects on two interrelated facets of household livelihoods: income and off-farm labour supply. They find that the NFPP has had a negative impact on incomes from timber harvesting, due to logging restrictions. However, they show that off-farm labor supply has increased more rapidly in NFPP areas than in non-NFPP areas. As such, the NFPP has actually had a positive impact on total household incomes from all sources. Furthermore, this result is strongest for employment outside the village. On the basis of these results, the authors also offer an intriguing discussion concerning how to mitigate the negative impact of the program and strengthen its positive impact on household livelihoods.

China's ecological restoration programs, which are sometimes referred to as "the Priority Forestry Programs," or PFPs, also include the Wildlife Conservation and the Nature Reserve Development Program (WCNR) and the Shelterbelt Development Program (SBDP). In addition to improving the environmental and resource conditions, a frequently reiterated goal of these PFPs is to enhance the income of rural residents. Thus, a question of common interest is: How has implementing the PFPs affected the farmers' income and poverty status? The task of Chapter 12 is to address this question, using a fixed-effects model and panel data from over 2,100 households in ten counties of Sichuan, Hebei, Shaanxi, and Jiangxi. The evidence indicates that their effects are mixed. The impacts of the SLCP, the NFPP, and the

DCBT are significantly positive, whereas the impact of the WCNR is negative and the SBDP has little effect on household income. Further, these impacts show substantial variations in different counties. Additionally, land for home gardening, labor for off-farm employment, and technical and institutional changes play major roles, indicating that more attention should be directed to program designing and income increasing from cash crops, off-farm employment, and education. Lastly, this chapter reminds us that when multiple ERPs are undertaken, focusing on one but ignoring the other(s) can result in incomplete or even biased findings.

Using the Malmquist index method and household survey data, the authors of Chapter 13 find that during the period of 1998–2004, the total factor productivity (TFP) in Wuqi of northern Shaanxi grew by 15.8%. By decomposing the TFP, they further show that its increase is due exclusively to the improvement of technical efficiency, rather than to technological change. To validate these findings and put them in perspective, they take a further step to derive the TFP change with county-level aggregate data. It is revealed that driven by technological change and scale efficiency, the TFP experienced a slow growth during the period of 1992–1998. Because of the tremendous cropland reduction and production mode shift caused by implementing the SLCP, the TFP declined substantially during the first three years of the program. Due to continued improvement of technical efficiency, however, its growth accelerated later. Altogether, their empirical evidence corroborate the fact that implementing the SLCP has contributed to the agricultural TFP growth in the longer term, and that the efficiency improvement has resulted mainly from the increased public expenditures for extension services and diffusion of technical knowledge. Wuqi's experience proves that it is possible to achieve environmental conservation and productivity increase simultaneously, even when facing a huge cropland reduction and production mode alternation.

In Chapter 14, the authors measure the aggregate socioeconomic impacts, direct and indirect, of the NFPP using input-output (I-O) models. They find that the NFPP will expand the annual output of the forest sectors by 5.8 billion yuan and the whole economy by 8.9 billion yuan by 2010. Employment will increase by 0.84 million in the forest sectors and by 0.93 million in the whole economy. Associated with the enormous expansion of forest protection and management are potential contributions to mitigating water runoff, soil erosion, flooding, and biodiversity loss. So they conclude that the investments and adjustments are worthwhile, if the program is properly implemented. In their view, the challenges are to transform loggers into tree planters and forest managers and ensure that the financial and institutional commitments by the local and national governments will be materialized.

Based on all of the empirical evidence of the environmental and socioeconomic impacts reported in this book, it seems clear that, by and large, China's ERPs have already had significantly positive impacts. This is so even with relatively short time-series data. Therefore, the government agencies, local farmers, business employees, policy practitioners, and other stakeholders should be encouraged by these affirmative findings. However, it should be pointed out that there are great opportunities to improve the effectiveness and impacts, and indeed, as insisted by many authors, a lot more can be done. This book is one of the only few publications of integrative assessment of the ERPs so far and further study of the socioeconomic impacts and particularly the ecological impacts should be pursued. For that matter, the science community needs to join hands in conducting quality assessment. This is truly an exciting and emerging area of research with great policy relevance and intellectual potential.

The book as a whole or its selective chapters can be used for teaching advanced classes on assessing the ecological and socioeconomic impacts of ecological restoration programs, evaluation of ecosystem services, and China's forestry and environmental conservation. The methodology, modeling frameworks, data-generating mechanisms, and analytic tools are useful to other researchers as well. In fact, some of the chapters have been published or accepted for publication as peer-reviewed articles, while others are being reviewed for publication in an *Environmental Management* special issue. It is my sincere hope, and indeed the hope of all of the contributors to this book, that its publication will promote more active assessment of the programs and more careful examination of the related policy measures to make them more effective.

East Lansing, MI   Runsheng Yin

Acknowledgments

First, I appreciate the contribution and cooperation of the authors of the individual chapters. Without their hard work and professionalism, this book would not have been possible. I am also grateful for the funding that my collaborators and I received from the US National Science Foundation (NSF), which has enabled us to conduct our research projects in China and thus to generate several manuscripts included in the book.

This book was based mostly on presentations made at the *International Symposium on Evaluating China's Ecological Restoration Programs*, jointly organized by the Forest Economics and Development Research Center (FEDRC) of the Chinese State Forestry Administration (SFA) and Michigan State University (MSU) in Beijing, on October 17, 2007. When I approached Dr. Lei Zhang and Dr. Guangcui Dai, Director and Deputy Director of the FEDRC, with the concept of organizing this event, they were very receptive and supportive. Similarly, when I brought my proposal for it to the relevant administrators and my colleagues at MSU, including Dr. Ian Gray, Vice President for Research and Graduate Studies; Dr. Jeff Armstrong, Dean of the College of Agriculture and Natural Resources; Dr. Jeff Riedinger, Dean of International Studies and Programs; Dr. Dan Keathley, Chairperson of the Department of Forestry; Dr. Robert Glew, Director of the Center for Advances Studies of International Development; and Dr. Weijun Zhao, Director of the Office of China Programs, they all indicated their encouragement to me as well as their willingness to sponsor the event. Later, Guiping Yin at the Ecosystem Policy Institute of China, my research office in Beijing, and Xiaojing Zhang at the FEDRC helped me prepare the symposium. Additionally, Dr. Kim Wilcox, MSU Provost, Dr. Steve Pueppke, Director of MSU Agricultural Experiment Station, Dr. Frank Fear, Senior Associate Dean of MSU College of Agriculture and Natural Resources, Dr. Michael Lewis, Director of MSU Asian Studies Center, and many other colleagues have supported my research in China in different ways. I am indebted to all of them.

When the symposium was held, keynote speeches were delivered by Mr. Yucai Li, SFA Deputy Administrator; Dr. Ian Gray, MSU Vice President for Research and Graduate Studies; Ms. Karin Wessman, country representative of the Worldwide Wildlife Fund; Professor Guofang Shen, member of the National Academy of Engineering and China Council for International Cooperation on Environment and Development; Dr. William Chang, Director of the US NSF Beijing Office and Science Attaché of the US Embassy in Beijing; and Mr. Yuhe Liu, former Deputy Minister of the Ministry of Forestry. Several project leaders from the SFA Department of Planning and Finance, Department of International Cooperation, program offices, and FEDRC reported their program implementation, monitoring, and preliminary assessment, and provided other assistance. Included in the participants were Jeff Riedinger, Dan Keathley, Wijun Zhao, Xiaowen Tang, Baohua Ye, Baoyu Li, Jianhua Bai, Chen Xie, David Newman, Bill Hyde, Andreas Kontolene, Emi Uchida, Shuqing Zhao, Tianjun Wen, Shiqu Zhang, Christopher Green, Schulz Wolfgaus, Naijun Dong, Junchang Liu, Can Liu, Shunbo Yao, Qin Xu, Xiufang Sun, Yueqin Shen, Qingshan Ren, and many others. Their contributions as invited speaker, presenter, panelist, session chair, or simply as attendee were gratifying to me.

Represented at the symposium were also such organizations as the Office of Policy Research, Office of Poverty Alleviation, and Development Research Center of the State Council, Environmental and Resource Legislation Office of the National People's Congress, Forestry Division of the State Development and Reform Commission, Ministry of Finance, People's University, Peking University, Beijing Forestry University, Zhejiang Forestry University, Nanjing Forestry University, Central-South Forestry University, Chinese Society of Forestry, Yunnan Department of Forestry, Shandong Institute of Forest Science, Journal of Forest Economics, Green China, and Green Times. Their presence and engagement made the symposium more relevant and successful. To these and other colleagues and friends, I would like to say "Thank you very much."

Dr. Fabio de Castro and Dr. Fritz Schmuhl, the two Publishing Editors of Environmental Science, and Mrs. Takeesha Moerland-Torpey, Editor of the Environmental Economics, Management, and Policy Program at Springer have been very patient and resourceful in helping me prepare the individual chapters and other material for printing. Upon my request, Dr. Roger Sedjo, Senior Fellow at Resources for the Future, agreed to write the Foreword for this book. I am sure that the readers will enjoy his concise articulation and expert insight. In addition, Lanying Li, Erin Shi, and Victoria Hoelzer-Maddox have provided valuable editing assistance. Finally, my wife, Xiu Du, and daughters, Fangfang and Anna, were inspirational and quietly supportive while I was editing this book. My special thanks go to all of them.

Contents

Contributors

Xiangzheng Deng, Center for Chinese Agricultural Policy, Institute of Geographic Sciences of Natural Resources Research, Chinese Academy of Sciences, Anwai, Beijing 100101, P.R. China, dengxz.ccap@gmail.com

Yulin Deng, College of Forestry and Horticulture, Sichuan Agricultural University, Ya'an 625014, P.R. China, yulindeng66@126.com

Zhaodong Feng, Department of Geology, Baylor University, Waco, TX 76798, USA; Key Laboratory of Western China's Environmental Systems (MOE), Lanzhou University, Lanzhou 730000, P.R. China, zhaodong_Feng@baylor.edu

Biyun Guo, Department of Forestry, Michigan State University, East Lansing, MI 48824, USA; College of Environment and Resources, Northwest A&F University, Yangling 712100, P.R. China, guobiyun@hotmail.com

Yajun Guo, College of Economics and Management, Northwest A&F University, Yangling, Shaanxi 712100, P.R. China, guoyajun71@126.com

Xuexi Huo, College of Economics and Management, Northwest A&F University, Yangling, Shaanxi 712100, P.R. China, dxhxx1988@yahoo.com.cn

Andreas Kontoleon, Department of Land Economy, University of Cambridge, Cambridge, CB3 9EP, UK, ak219@cam.ac.uk

Changbin Li, Key Laboratory of Western China's Environmental Systems (MOE), Lanzhou University, Lanzhou 730000, P.R. China; Center for Global Change and Earth Observations, Michigan State University, East Lansing, MI 48823, USA, licb0701@gmail.com

Hua Li, College of Economics and Management, Northwest A&F University, Yangling, Shaanxi 712100, P.R. China, lihua7485@163.com

Lanying Li, School of Economics and Management, Zhejiang Forestry University, Lin'an 311300, P.R. China, llycds@msu.edu

Zhengpeng Li, ASRC Research and Technology Solutions, Contractor to USGS EROS Center, Sioux Falls, SD 57198, USA, zli@usgs.gov

Xianchun Liao, Department of Forestry, Michigan State University, East Lansing, MI 48823, USA, liaoxian@msu.edu

Can Liu, China National Forestry Economics and Development Research Center, State Forestry Administration, Beijing 100714, P.R. China, liucan@public.bta.net.cn

Guangquan Liu, Chinese Academy of Water Resources and Hydropower Research, Beijing 100044, P.R. China, gqliu@iwhr.com

Shuguang Liu, US Geological Survey (USGS) Earth Resources Observation and Science (EROS) Center, Sioux Falls, SD 57198, USA, sliu@usgs.gov

Jinzhi Lü, China National Forestry Economics and Development Research Center, State Forestry Administration, Beijing 100714, P.R. China, lv_jinzhi@yahoo.com.cn

Katrina Mullan, University of Cambridge, Department of Land Economy, Cambridge CB3 9EP, UK, klm31@hermes.cam.ac.uk

Jiaguo Qi, Center for Global Change and Earth Observations, Michigan State University, East Lansing, MI 48823, USA, qi@msu.edu

David Rothstein, Department of Forestry, Michigan State University, East Lansing, MI 48824, USA, rothste2@msu.edu

Scott Rozelle, Freeman Spogli Institute, Food Security and Environmental Program, Stanford University, Stanford, CA 94305-6055, USA, rozelle@stanford.edu

Yueqin Shen, School of Economics and Management, Zhejiang Forestry University, Lin'an 311300, P.R. China, shenyueqin@zjfc.edu.cn

Zhangquan Shen, College of Environmental and Resource Sciences, Zhejiang University, Hangzhou 310029, P.R. China; Center for Global Change and Earth Observations, Michigan State University, East Lansing, MI48823, USA, zhqshen@zju.edu.cn

Tim Swanson, University College London, Department of Economics and School of Laws, London, WC1E, UK, tim.swanson@ucl.ac.uk

Kun Tan, Department of Ecology, Peking University, Beijing 100871, P.R. China, tank@pku.edu.cn

Emi Uchida, Department of Environmental and Natural Resource Economics, University of Rhode Island, Kingston, RI 02881, USA, emi@uri.edu

Qing Xiang, Department of Forestry, Michigan State University, East Lansing, MI 48824, USA, xiangqin@msu.edu

Jintao Xu, College of Environmental Sciences, Peking University, Beijing 100871, P.R. China, xujt@pku.edu.cn

Shunbo Yao, College of Economics and Management, Northwest A&F University, Yangling, Shaanxi 712100, P.R. China, yaoshunbo@126.com

Guiping Yin, Ecosystem Policy Institute of China, Building A, Chaoyang District, Beijing 100102, P.R. China, epicyin@yahoo.cn

Runsheng Yin, Department of Forestry, Michigan State University, East Lansing, MI 48824, USA; Ecosystem Policy Institute of China, Chaoyang District, Beijing 100102, P.R. China, yinr@msu.edu

Feng Zhang, Center for Global Change and Earth Observations, Michigan State University, East Lansing, MI 48823, USA; Key Laboratory of Arid and Grassland Agro-Ecology (MOE), Lanzhou University, Lanzhou 730000, P.R. China, zhangfeng04@gmail.com

Shiqiu Zhang, College of Environmental Sciences, Peking University, Old Earth Science Building, Beijing 100871, P.R. China, zhangshq@pku.edu.cn

Shuqing Zhao, Department of Forestry, Michigan State University, East Lansing, MI 48824, U.S.A.; Center for Global Change and Earth Observations, Michigan State University, East Lansing, MI 48823, USA, szhao@usgs.gov

List of Figures

List of Tables

Chapter 1
China's Ecological Restoration Programs: Initiation, Implementation, and Challenges

Runsheng Yin and Guiping Yin

Abstract China has been undertaking several major ecological restoration programs in recent years, including the Sloping Land Conversion Program, the Natural Forest Protection Program, and the Desertification Combating Program around Beijing and Tianjing. This chapter summarizes how these programs have been initiated and implemented, and what the main challenges are in carrying them forward. It seems that with huge government investments, tremendous progress has been made in implementing them. However, in order to complete them successfully and to fundamentally improve the ecosystem functions and services, it is essential for China to have a more balanced and comprehensive approach to ecological restoration; adopt better planning and management practices; strengthen the governance of program implementation; emphasize local people's active engagement; establish an independent, competent monitoring network; and conduct timely and high-quality assessments of the program effectiveness and impacts. We believe that in each of these areas, the international community can and should provide a wide range of technical assistance.

Keywords Land conversion · Forest protection · Desertification · Program effectiveness and impact · Policy improvement

1.1 Introduction

Induced by population pressure, economic growth, and historic exploitation, a large portion of China's primary forests and wetland was depleted, and a high percentage of its farmland and grassland was degraded (WWF, 2003; Yin et al., 2005). These ecosystem disturbances caused extensive desertification, flooding, soil erosion, dust storms, elevated levels of greenhouse gas emissions, and severe damage to wildlife habitat (Liu & Diamond, 2005; Xu, Yin, Li, & Liu, 2006). To address these

R. Yin (✉)
Department of Forestry, Michigan State University, East Lansing, MI 48824, USA;
Ecosystem Policy Institute of China, Chaoyang District, Beijing 100102, P.R. China

R. Yin (ed.), *An Integrated Assessment of China's Ecological Restoration Programs*,
DOI 10.1007/978-90-481-2655-2_1, © Springer Science+Business Media B.V. 2009

concerns and to improve its environmental conditions, China has been undertaking several major ecological restoration efforts (EREs), including the Natural Forest Protection Program (NFPP), the Sloping Land Conversion Program (SLCP), and the Desertification Combating Program around Beijing and Tianjin (DCBT). These EREs represent large-scale, transformational changes and will have profound environmental and socioeconomic impacts both domestically and internationally (Forest and Grassland Taskforce of China, 2003; Wang, Innes, Lei, Dai, & Wu, 2007).

A few studies have investigated the early implementation of these programs, their preliminary impacts, and the perceived problems in carrying them forward. For instance, Xu et al. (2006) was one of the earliest efforts that summarize the essentials, progress, and effects of the NFPP and SLCP. The authors also commented on the challenges facing the programs. Similarly, Liu et al. (2008) reported the implementation of these two programs, reviewed the patchy evidence of their effects, and suggested steps for improving their effectiveness, including systematic planning, diversified funding, effective compensation, integrated research, and comprehensive monitoring. In addition, Wang, Innes, et al. (2007) presented a synopsis of these and other programs and then related them to the latest forest tenure reform and other policy initiatives of the Chinese government. Undoubtedly, China's experiences and lessons of ecological restoration are of broad interest and a lot has been learned of the NFPP and the SLCP from these and other studies.

Nonetheless, China's EREs are not limited to the NFPP and the SLCP (Wang, Innes, et al., 2007; Liu et al., 2008), and a complete documentation and a timely updating of all the major efforts are still missing in the literature, which are not conducive to gauging the scope of these programs and the scale of their impacts. Additionally, a more thorough and critical deliberation of the relevant policy and technical measures remains urgently needed for improving the implementation of these programs. The purpose of this paper is to tackle these tasks and thereby to advance the international understanding and to facilitate the execution of China's EREs. Before proceeding, it should be noted that because of space limitation and potential diffusion of attention, we have decided not to review the literature that has assessed the socioeconomic and ecological impacts of the EREs here.[1] Furthermore, given that our tasks are to narrate the basic contents of the programs and their progress and to discuss the major issues encountered in their execution, a number of the statistics will be drawn from official sources and a largely descriptive analysis will suffice.

The paper is organized as follows: In the remainder of this section, we will provide the historic background, against which China's EREs were launched; then, we will present their key elements and implementation in the next two sections; finally, we will address the critical issues that China faces in successfully completing these programs and fundamentally improving its ecosystem functions and services.

After the People's Republic of China was established in 1949, large tracts of primary natural forests remained in the northeast, southwest, and a few other places.

---

[1] Readers interested in this literature can refer to Chapter 2 for a comprehensive review.

Later, most of these forests were nationalized and 136 state-owned forest bureaus were gradually set up in these forests to produce timber to spur the young economy (SFA, 2001). Along with this strategy of resource exploitation, the old governance system came into play, under which the state enterprises lacked incentive and autonomy to manage and utilize the resources efficiently (Yin, 1998). Since logging was the main, or even the sole, revenue source and the forest bureaus had to assume the heavy burden of providing almost all of the social services for their workforce, over-cutting became prevalent and regeneration and management were neglected. At the same time, population growth and demand for employment in these forest regions led to more fuelwood consumption, housing construction, and land clearing. Consequently, China's natural forests were quickly depleted (Liu, 2002). The over-cutting and under-management also resulted in structural deterioration of the forests, as reflected in reduced stocking volume, imbalanced age structure, altered species composition, and low growth rate (Yin, 1998).

Unfortunately, the ecological environment in the rural society was even worse. The collectivization in the 1950s discouraged people from tree-planting and forest management. Soon after, as part of the attempt to industrialize, a campaign to increase steel production took place during the Great Leap Forward. In many cases, even backyard furnaces fueled by wood charcoal were deployed for that purpose (Yin, 1994). Throughout the 1960s and 1970s, forest and grassland conversion to other uses was also carried out under the chronic crisis of grain shortages on the one hand and the national policy of food self-sufficiency on the other. These malpractices resulted in severe destruction of vegetation and deforestation. Also, farming on steep slopes became common due to demographic explosion and lapses in regulation (Du, 2001). Coupled with uneven rainfalls and rugged terrains, this led to a substantially reduced ecosystem capacity to regulate water flows and to contain soil erosion (Lu et al., 2002). These factors were deemed to be the primary reasons for the record dry-up of the Yellow River in 1997 and the widespread flooding in the Yangtze basin in 1998 (Xu & Cao, 2001). Another disaster has been the loss of grass cover and desertification in the west, driven by uncontrolled grazing, poor maintenance of rangeland, and human-induced decline of water tables (Yin et al., 2005).

To be sure, the Chinese government made attempts to combat the growing environmental problems in the past, but the record of their effectiveness was utterly disappointing. Since 1978, for instance, a number of afforestation projects has been launched (Xu et al., 2006). Despite their broad geographic reach and remarkable planting efforts, public investments were limited, and the efforts were rarely followed through (Smil, 1993). Often, sites were poorly prepared, seeds and/or seedlings were not properly planted, and saplings were not well tended. Therefore, these forestry projects have failed miserably in delivering the expected environmental benefits. Similar problems have hindered the efforts to curb farming on slopes (Xu & Cao, 2001). In addition, having gained economic independence from the rural reforms, many farmers aggressively sought new croplands, just like herders hastily increased their livestock (Du, 2001). Oftentimes, this meant that a large number of the more sensitive patches on steeper slopes were claimed, and a great amount of the grassland in the arid and semi-arid regions was overgrazed.

In short, the successive occurrences of ecological disasters in the late 1990s indicated that China's historic efforts to protect ecosystems had not improved the overall ecological conditions and more decisive and forceful measures are thus called for to bring the problems under control. It was against this backdrop that the new EREs were launched and the existing ones consolidated. The State Forestry Administration (hereafter, SFA), prompted by the ecological disasters and charged by the State Council, proposed the NFPP in 1998 as a large-scale scheme to protect most of the natural forests and, then, the SCLP in 1999 to convert croplands on slopes and desertified fields back to forestland, grassland, and wetland on an even larger scale.[2]

Moreover, China has indeed been carrying out four other programs: the Desertification Combating Program around Beijing and Tianjin, the Shelterbelt Network Development Program, the Wildlife Conservation and Nature Reserves Protection Program, and the Industrial Timberland Plantation Program (SFA, 2002). Altogether, they are designated by the SFA as the "Six Priority Forestry Programs" of ecological restoration and resource expansion and are incorporated into the national economic development and environmental protection plans (SFA, 2007; Wang, Innes, et al., 2007).[3] The Chinese government hopes that these programs will not only greatly improve the domestic resource and ecological conditions as well as rural livelihoods, but also significantly contribute to regional and global environmental causes (Yin et al., 2005; Liu et al., 2008). The total investment in these programs over this decade will easily top 500 billion yuan (Wang, Innes, et al., 2007).[4]

1.2 Program Contents

1.2.1 The NFPP

Because a dominant portion of the natural forests is national forests managed by the state forest enterprises (Yin, 1998), the NFPP is thus geared towards national forests and state forest enterprises. After 2 years of trial, it was formally launched in 2000. The specific goals of the NFPP are to: (1) reduce timber harvests from natural forests from 32 million m^3 in 1997 to 12 million m^3 by 2003; (2) conserve nearly 90 million ha of natural forests; and (3) afforest and re-vegetate an additional 8.7 million ha by 2010 by means of mountain closure, aerial seeding, and artificial planting (Liu, 2002).

---

[2]The SLCP is also known as the "Grain for Green" in the literature (WWF, 2003; Xu et al., 2006).

[3]They are called forestry programs because they have a clear forest orientation and are managed by the State Forestry Administration. Notably, there have been other ecological restoration programs, such as the water and soil conservation ones undertaken by the Ministry of Water Resources and the farmland and grassland protection ones administered by the Ministry of Agriculture. However, the geographic coverage and public investment in these programs are generally much smaller.

[4]This is equivalent to roughly US\$70 billion given the current exchange of \$1 = 6.85 yuan.

As a result of the logging bans and harvest reductions, 740,000 loggers and other workers in the downstream of the timber supply chain were displaced. So, the NFPP also stipulates that those displaced employees would be transferred to forestation and forest management activities, retired, or laid off, depending on their individual status and choice, and reemployment opportunities, among other things. In addition, all existing and newly retired employees would be incorporated into the provincial pension and social security systems; for the laid-off workers, a minimum living expenditure would be assured. This could have created a great budgetary burden for the provincial and county governments where forestry was a main sector. To alleviate the problem and explore new business opportunities, the financial burden is partly shared by the central government.

The widespread defaults of the state forest enterprises on their financial obligations caused by the drastic business disruption also mean that the government would have to write off a large amount of business debt. Additionally, to protect existing resources, expand resource coverage, and improve resource quality, the government has allocated a large sum of funds for the conceived forestation, re-vegetation, and land management activities. The original budget of the program was 96.4 billion yuan until 2010. Of that total, 85% would be covered by the central government and 15% by governments of the involved provinces (SFA, 2001).

1.2.2 The SLCP

The land conversion piloting began in 1999 in Sichuan, Shaanxi, and Gansu, located in the middle and upper reaches of the Yellow River and the upper reaches of the Yangtze River. The SLCP was originally planned for the period of 2001–2010 in two phases. The first phase, from 2001 to 2005, was aimed at a preliminary control of the fragile ecological situation in the program areas. A main task was to retire and convert 11.33 million ha of sloping farmlands, including 4.4 million ha on slopes steeper than 25°, and desertified fields. Meanwhile, another 13.33 million ha of sparsely vegetated mountainous, hilly, and sandy lands was included for forestation or re-vegetation. It was envisioned that these steps would lead to not only a significant increase of vegetation cover, but also erosion control and desertification curtailment on a large scale. During the second phase, from 2006 and 2010, the SLCP was designed to retire and convert an additional 3.3 million ha of farmlands and afforest and re-vegetate 4.0 million ha more of the sparsely vegetated hilly and sandy fields. It was expected that by 2010, the vegetation cover would be further expanded, ultimately resulting in erosion control over 86.67 million ha and desertification containment over 102.67 million ha (SFA, 2002).

The SLCP mandates that farmers who participate in the land retirement and conversion be compensated. The retired farmland may be converted into ecological forests (forests primarily providing ecological functions and services), commercial forests (forests producing timber, fruits, nuts, and medical and other commodities), or grass cover as appropriate. The compensation scheme includes annual grain and cash subsidies, and free seeds/seedlings at the beginning of the conversion. The subsidies are for 8 years if ecological forests are established, 5 years if economic forests are established, and 2 years if grass cover is established (Xu & Cao, 2001). To account for regional variations of crop yields and population densities, the grain subsidy is set at 2,250 kg/ha in the Yangtze River basin and 1,500 kg/ha in the Yellow River basin. The cash subsidy is 300 yuan/ha for eligible land each year. Finally, agricultural taxes on the converted lands are exempted and the loss of local revenues due to reduced agricultural output and shrunk tax base are shared by the central government. The total investment was projected to be 225 billion yuan for this decade (Tang, 2007).

1.2.3 The DCBT

In the spring of 2000, sandstorms invaded northern China twelve times. The high frequency, broad coverage, and large damage of these events had seldom been seen over the past half century. Shocked by these events, the central government quickly drew up a blueprint to inhibit the encroachment of desertification and to improve the natural environment for Beijing, the capital of China, and its adjacent areas, thus leading to the commission of the DCBT. The program was also drawn up for the period of 2001–2010. The main target of the DCBT is to treat 10.13 million ha of desertified lands, of which 5.21 million ha is to be vegetated or re-vegetated. To fulfill the tasks, adopted measures include conversion of cropland to forests, grassland rehabilitation, selective banning of open grazing, integrated watershed management, and ecological resettlement.

The area of cropland that would be converted to forest and grass coverage is roughly 2.63 million ha. Meanwhile, over 4.94 million ha of degraded, sandy fields will be afforested or reforested, 10.63 million ha of grassland rehabilitated, 23,445 km^2 of small watersheds protected, and 180,000 villagers resettled.[5] Farmers and herders involved in these activities are compensated as well. In addition to cropland retirement, which receives a subsidy similar to that under the SLCP, compensations for other activities are the following: afforestation at 4,500 yuan/ha, forest regeneration via aerial seeding at 3,000 yuan/ha and mountain closure at 1,050 yuan/ha, grassland establishment via artificial planting and aerial seeding at 1,800 and 1,500 yuan/ha, grassland fencing at 1,050 yuan/ha, livestock pen building at 200 yuan/m^2, integrated watershed management at 200,000/km^2, and ecological resettlement at 5,000 yuan/person. The program covers 75 counties in Beijing, Tianjing, Hebei, Shanxi, and Inner Mongolia. The estimated total investment is 57.7 billion yuan (SFA, 2003).

---

[5]In certain places, the carrying capacity of the degraded farmland and grassland ecosystems has become so abysmal that the farmers and herders have no choice but to get resettled to other more viable locations. Understandably, the degraded lands will be covered in the rehabilitation effort in this case. The question remains, though, how to guarantee the farmland and grassland around the newly resettled sites will not be degraded in the future.

1.2.4 Other Programs

In this section, we synthesize the other three EREs – the Wildlife Conservation and Nature Reserves Protection Program (WCNR), the Shelterbelt Network Development Program (SNDP), and the Industrial Timberland Plantation Program (ITPP).

The WCNR strives to expand the number of nature reserves and enhance the protection of wildlife. It sanctions that the total number of reserves should reach 1,800 by 2010 and 2,000 by 2030; and it stipulates that the primary protected areas be administrated by the central and provincial governments, while smaller and less critical areas managed by the municipal and county governments (SFA, 2002). In addition, included in it are measures of wetland restoration, ecotourism development, wildlife breeding, as well as strengthening the role of science and technology, particularly monitoring and evaluation of reserves and biodiversity. The total planned investment is 135.65 billion yuan over a period of 30 years from 2001, with roughly a half (66.5 billion) by the central government. The program also seeks active domestic and international participation, including broad involvement of the private sector.

Covering the vast Three Norths (northwest, north, and northeast), the Yangtze River basin, the Zhujiang River basin, and the Taihang Mountain Range, the SNDP intends to mobilize public agencies, civil society, and individuals to engage in shelterbelt development and maintenance and extensive tree planting.[6] Its goal is to mitigate wind-induced erosion, landslides, and flooding, and to protect grasslands, riverbanks, and coastal lines (SFA, 2002). With limited investment by the central government – 70 billion yuan over the period of 2001–2010, however, it aggressively seeks regional investments and local labor contributions. Additionally, it supports the adoption of appropriate silvicultural techniques and the integration of shelterbelts with farming and grazing by means of agroforestry practices.

The ITPP represents a major market-driven effort for increasing domestic timber supply (SFA, 2002). To induce private initiative and engagement, as high as 70% of the investment may come from loans subsidized by the National Development Bank and tax incentives are prescribed as well. It also urges active involvement by various types of business entities – state or collectively owned, shareholder based, or fully private. The planned area of establishment is 4.7 million ha from 2001 to 2005, 9.2 million ha by 2010, and 13.3 million ha by 2015, respectively. The projected total investment by the government is 71.8 billion yuan.

The six programs are summarized in Table 1.1. Obviously, these EREs vary hugely in terms of their mission, time coverage/duration, and financial commitment, among other things. Nonetheless, the SLCP is by far the largest. Notably, except for the SLCP and the NFPP, none of these programs were brand new or recently initiated, as some authors have suggested. Rather, certain forms of their predecessors

---

[6]The shelterbelt network in the Three Norths, once fully developed, will constitute a vast belt of 400–700 km wide and 4500 km long that parallels the Great Wall. Thus, it is often called the "Green Great Wall" by the government and the media (SFA, 2002).

Table 1.1 Key policy measures of the Six Priority Forestry Programs

Program	Key policies
Sloping Land Conversion Program (SLCP), covering 25 provinces during 2001–2010	• Sloping or desertified cropland is converted into ecological/economic forest, and grassland; ecological forest should account for 80% of the converted land. • The central government subsidizes farmers with seeds or seedlings, grain, and cash. • Subsidies last 8 years for ecological forest, 5 years for economic forest, and 2 years for grassland. The annual cash subsidy is 300 yuan/ha, and the annual grain subsidy is 1,500 kg/ha in the Yellow River basin and 2,250 kg in the Yangtze River basin. • The central government also makes fiscal transfers to compensate the entailed losses to local fiscal revenues. • The estimated total investment is 225 billion yuan.
Natural Forest Protection Program (NFPP), covering 17 provinces during 2000–2010	• Complete ban on commercial logging in the upper Yangtze and upper and middle Yellow River basins and sharp reduction in commercial harvests in other regions. • Shutting down certain processing facilities, compensating logging firms, and dealing with displaced workers and equipment. • Promotion of afforestation and forest management wherever necessary. • Strengthening administration and law enforcement, including forest protection. • Restructuring the forest industry, and improving the efficiency of timber utilization. • The initial investment commitment is 96.4 billion yuan.
Wildlife Conservation and Nature Reserve Development Program (WCNR), scattered all over the country during 2001–2050	• Priority protected areas are administrated by the central government, while smaller and less critical areas are managed by the regional governments. • Established reserves will reach 1,800 by 2010, 2,000 by 2030, and 2,500 by 2050. • Included are also wetland protection and restoration, ecotourism development, and wildlife breeding. • Encouraging domestic and international participation and contributions, including broad involvement of the private sector. • Strengthening the role of science and technology, particularly nature reserve and biodiversity monitoring and evaluation. • Total planned investment is 135.65 billion yuan, with roughly a half covered by the central government.
Shelterbelt Development Program (SBDP), covering all 31 provinces during 2001–2010	• Including the Three Norths (northwest, north, and northeast), the Yangtze River basin, the Zhujiang River basin, and the Taihang Mountain Range. • Mobilization of public agencies, civil society, and individuals to participate in shelterbelt development and tree planting. • Encouraging regional government investment and local labor contribution, and adopting new silvicultural techniques. • Total planned investment is 70 billion yuan.

Table 1.1 (continued)

Program	Key policies
Desertification Combating around Beijing and Tianjing (DCBT), including Inner Mongolia, Hebei, Shanxi, Beijing, and Tianjin during 2001–2010	• Converting desertified land into forestland and grassland by means of flexible and diversified measures based on the local conditions. • Changing herding and animal husbandry practices to control overgrazing and rehabilitate degraded grassland. • Developing irrigation projects and resettling people away from fragile areas. • Extension of suitable production technology and energy sources. • Establishing desertification monitoring and dust storm forecasting systems. • Total projected investment is 57.7 billion yuan.
Industrial Timber Plantation Development Program (ITPP), covering 18 provinces during 2001–2015	• Market-driven and profit-orientated efforts for increasing domestic timber supply. • As high as 70% of the investment may come from loans subsidized by the National Development Bank. • Tax incentives are provided. • Encouraging active participation by various enterprises – state or collectively owned, shareholder based, or fully private. • Planned area of establishment is 4.69 million ha by 2005, 9.2 million ha by 2010, and 13.33 million ha by 2015. • Projected total investment is 71.8 billion yuan.

had been in existence much earlier and they were consolidated a few years ago for more effective administration or reoriented for more focused targeting (SFA, 2002). For example, the shelterbelt expansion in the Three Norths, the upper Yangtze basin, and other regions were originally launched in the 1970s and 1980s. It was only in 2001 when they were combined under the single SNDP umbrella. Another example is desertification combating, which was and still is carried out on a vast geographic scale. The DCBT was an outgrowth and thus a key component of this undertaking, in response to the frequent invasion of dust storms to Beijing, Tianjin, and other cities in the northern plains (SFA, 2002). And desertification combating in other areas now overlaps with and even has become part of the SLCP.

1.3 Implementation and Outcome

Since their initiation, farmers, herders, state employees, and other stakeholders as well as communities and public agencies have enthusiastically participated in these ambitious EREs. It appears that with the unprecedented government financing, remarkable progress has been made in implementing them (Xu et al., 2006; Wang, Innes, et al., 2007; Liu et al., 2008). Therefore, the government claims that the ecosystem conditions and people's livelihoods have been significantly improved (SFA, 2007).

1.3.1 Implementation

In carrying out the NFPP, the targets of logging bans, commercial harvest reduction, resource protection, reforestation, and afforestation have been well met. Timber production has been restricted to the quota determined by the State Council. In 2006, for instance, the total production of commercial roundwood was 13.5 million m^3 in the whole program area. Over 90 million ha of natural forests has been effectively protected, and forestation via artificial planting, aerial seeding, and mountain closure has reached 15 million ha. Regional and local authorities have made concerted efforts to develop alternative business opportunities and transform the forest-based economy. Ecotourism and other activities – dairy, cattle, and deer farming, growing annual crops, mushrooms, fruits, and ginseng, and collecting wild herbs, nuts, and vegetables – have gained broad recognition. Also, more and more of these activities have been undertaken by private enterprises.

Meanwhile, 665,000 state employees in logging, hauling, and wood products processing have been terminated, with over 200,000 transferred to forest protection and management and 353,600 terminated their contracts (SFA, 2007). A social safety net, including employee pension and medical and unemployment insurance, has been put in place to cover 99% of employed and retired personnel. Statistics suggest that a bulk of the investment (68% prior to 2004) in state forest bureaus has been spent on employee settlement and benefits, rather than on actual forest protection, management, and tree planting (SFA, 2006). Thus far, a total of 57.5 billion yuan has been invested, of which 53.9 billion yuan has been made by the central government, accounting for 93.69%. Implementing the NFPP has also caused a large amount of the logging, hauling, and processing assets in state forest enterprises to become obsolete. To get these assets disposed and to write off the principals and interests of loans acquired by these firms, the central government has added 24.5 billion yuan to its budget (Tang, 2007). Hence, the actual investment is expected to be no less than 120.7 billion yuan in this decade. And there have been indications that the central government will extend the program well into the next decade (SFA, 2008).

In March 2000, the SLCP enrolled 174 counties in Yunnan, Guizhou, Sichuan, Hubei, Shanxi, Henan, Shaanxi, Gansu, Ningxia, Qinghai, Xinjiang, and Chongqing. Then, in 2001, it was expanded to 20 provinces, 400 counties, and 27,000 villages. Participation grew to 120 million farmers in 32.5 million households of 2,291 counties, with a concentration in the west (SFA, 2006). In the first 3 years, 1.2 million ha was converted, with an expenditure of 3.65 billion yuan (Xu & Cao, 2001). In 2002 and 2003, however, converted cropland jumped to 2.9 million ha and 3.3 million ha, respectively (SFA, 2004). By the end of 2004, a total of about 9 million ha of cropland had been retired, including 4 million ha on slopes steeper than $25°$. But thereafter, the government scaled back the cropland retirement abruptly – restricting annual enrollment within 0.67 million ha, due to the reality of dwindling grain surplus and concern for food security (Xu et al., 2006). These unexpected ups and downs and the added issues of set-aside enrollment and site and seedling preparation have made it difficult to execute the program smoothly. Meanwhile, the subsidies have shifted to the provision of seeds/seedlings and payment made wholly in cash, with grain valued at a fixed rate of 1.4 yuan/kg (Tang, 2007).

From the onset of the SLCP, the central government promised that the period of grain and cash subsidies would be extended if, after review, there were qualified lands (Du, 2001). Nevertheless, there had been a general concern regarding whether or not the government would honor its promise. In 2007, the State Council announced that the program would be extended until 2021, considering that many participating farmers are still facing poverty and having difficulty of finding alternative employment and income opportunities after having converted their farmland (Tang, 2007). However, the new scheme has begun with a major adjustment: In addition to keeping the original cash subsidy of 300 yuan/ha/year, the amount of subsidized grain has been halved. Consequently, the total subsidy has become 1,875 yuan/ha/year in the Yangtze River basin and 1,350 yuan/ha/year in the Yellow River basin. Again, the subsidies will last 8 years if ecological forests are planted, 5 years if economic forests are planted, and 2 years if grass cover is established.

With 206.6 billion yuan added to it according to the adjusted policy, the total investment of the SLCP up to 2021 will amount to 431.1 billion yuan (Tang, 2007). It was also indicated that a special fund would be set up by the Ministry of Finance to promote the long-term development of farmers' livelihoods and environmental recovery. This means that there will be continued investment in poverty alleviation, infrastructure expansion, ecological resettlement, alternative energy, and maintenance of forests and grassland (Tang, 2007).

Progress of the DCBT is also impressive. Up to 2006, afforestation expanded to 2.8 million ha by means of artificial plantation and aerial seeding, and to 1.3 million ha by means of mountain closure. Also, about 1.5 million ha of grassland was rehabilitated, 0.5 million ha of small watersheds treated, and over 101,000 poverty-stricken people resettled to places where a basic livelihood can be sustained (SFA, 2007). Total investment in the program reached to 13.6 billion yuan from its inception in 2000 to 2006. Surveys show that most of the investment has been used for subsidized cropland conversion to forests and grassland.

Likewise, two other programs – the WCNR and the SNDP – have made tremendous advances. The number and area of nature reserves have increased steadily, and the conservation of wild plants and animals has been enhanced greatly. By the end of 2006, there were already 1,740 nature reserves, accounting for 12.6% of the country's land base.[7] Through these reserves, 90% of the terrestrial eco-zones, 85% of the wild animal species, 65% of the plant communities, and 45% of the wetland have been put under protection (SFA, 2007).[8] Similarly, during the period of 2001–2006, afforestation, reforestation, and other land rehabilitation schemes were widely executed under the SNDP, which gained 3.25 million ha in the Three Norths

---

[7] Of the 1740 nature reserves, 198 are national ones, 583 provincial ones, 289 municipal ones, and 670 county-level ones (SFA, 2006).

[8] Included in them are over 300 rare and endangered animal species, including panda, tiger, elephant, monkey, and crane, as well as more than 130 rare and endangered plant and tree species (SFA, 2006).

and 3.05 million ha in the Yangtze River basin (SFA, 2007). In comparison, less than 0.4 million ha of industrial timber plantations were established during the same period under the ITPP (SFA, 2007), suggesting its very slow development.

1.3.2 Preliminary Outcomes

According to government reports, the natural resource conditions have been much improved due to implementing the EREs. During the period of 1999–2006, the forest area gained 8.1 million ha, and the stocking volume increased by 466 million m^3 in areas covered by the NFPP (Tang, 2007). Similarly, a large amount of degraded farmland and grassland has been converted and rehabilitated, and the forest and grassland coverage expanded substantially in implementing other programs (SFA, 2007). As a result, the ecological conditions have been improved as well, as broadly reflected in the decline of soil erosion and water runoffs, expansion of wildlife habit and species abundance, and reduction of sandstorm and flood occurrence. At the same time, the economic structure has been fundamentally adjusted, as indicated by the flourishing of non-timber forest products, growth of ecotourism and recreation, diversification of local economies, acceleration of labor transfer, increase of income and living standards, and reduction of poverty incidence (SFA, 2007).

While these statements are generally true, where, how, and to what extent these programs have changed the ecological and socioeconomic conditions remains poorly understood. Of course, this is partly because certain government statistics may not be available or reliable, and partly because the government has not given adequate attention to program impact monitoring and assessment. For instance, the SFA (2006) reported that the desertified land decreased 6,416 km^2 from 1999 to 2004; as such, the intensity of wind erosion has been weakened, and sandstorms and days of strong winds have been reduced in Beijing and its vicinity. However, these findings were not independently derived or even verified by the science community. Further, several analyses of selected sites in Hebei, Inner Mongolia, Gansu, and Qinghai suggest that the DCBT and other similar undertakings had rarely shown a significantly positive effect until 2005 (e.g., Zan & Wang, 2006; Wei et al., 2006; Yin 2007). In another case, the government asserts that the marked decline of erosion and sediment in the upper Yangtze basin, as reflected in the records of some hydrologic stations, is due largely to the combined effects of the SLCP and the NFPP. In Sichuan province alone, implementing these programs has led to a sediment reduction of 53 million tons and an increase of ecosystem water retention by 684 million tons a year (SFA, 2006). Again, our research shows that so many dams and reservoirs have been built in the region over the past two decades that their holding capacity has increased steadily (Yin, 2007). It is far from clear whether and how much the reduction of erosion and sediment is due to ecological restoration efforts or hydro-engineering projects. Therefore, much more efforts need to be made in strengthening the monitoring and assessment of the programs.

Actually, Xu et al. (2006), Uchida et al. (2007), and Liu et al. (2008) are among the few studies in the international literature that have shown evidence corroborating certain official claims of the program impacts. Still, their evidence is preliminary as well as sketchy; and some of it came from second-hand sources. Because the literature of impact assessment will be reviewed separately (see Chapter 2) and several studies evaluating the specific ecological and/or socioeconomic effects of China's EREs are included in this book, here it suffices to point out the salient gaps of program impacts between the government claims and the assessed outcomes.

1.4 Main Challenges

Despite the admirable intentions, huge investments, and seemingly tremendous achievements, China's EREs have faced huge challenges. Most, if not all, of them have been alluded to in the previous sections – the inadequacy of program monitoring and assessment, the heavy reliance on state financing, the rigidity and inconsistency of certain policy measures, the lack of inter-agency cooperation and careful planning, the insufficient consideration of local interests, and the neglect of appropriate technical practices. Understandably, these challenges are intertwined. Below, we elaborate them in detail and suggest ways for overcoming them. Interested readers can refer to Xu et al. (2006) or Liu et al. (2008) for an early discussion.

1.4.1 A More Balanced and Adequate Approach to Ecological Restoration

First, in contrast to the fervor and enthusiasm in planting trees, the SFA has shown less interest in other measures in ecological restoration, even if they are better suited in certain circumstances. Back in 2003, the Forest and Grassland Taskforce already pointed out that "Implementation has not been tailored to local conditions, and there has been an overemphasis on tree planting rather than restoring original vegetation cover." More recently, several commentators have voiced their concern that planting tall trees in semiarid and arid northwestern regions may not work well for the environment (Normile, 2007; Wang, Ouyang, et al., 2007; Cao, 2008). They argue that planting poplars, as a major species for afforestation, in those regions is problematic given the limited precipitation. In many cases, it is difficult to get the trees established; and wherever they are established, their deep root system can hemorrhage ground water through transpiration, lowering the water table and making it harder for native grass and shrubs to survive (Normile, 2007; Wang, Lu, Fang, & Shen, 2007). So, the time has come for the SFA to carefully evaluate its policy of targeting and species and site selection.

Indeed, one may add that even the forest management activities ensuing tree planting and regeneration, such as thinning and tending, have not been well incorporated into the programs. And this issue has been confounded by the high initial planting densities driven by the general requirement for high survival rates (Wang, Ouyang, et al., 2007; Yin et al., 2005). As a result, the growth rate tends to be low after canopy closure, and the forest quality, let alone the ecosystem functionality, has not been very satisfactory. This will in turn make the forests vulnerable to fire and pest attacks. More attention should thus be paid to the quality of both forest establishment and management. The historic lesson of large areas of afforestation but poor productivity of forests must not be forgotten (Yin, 1998).

To strengthen forest management, better technical work and broader support are called for. Regarding the former, improvements can be made in site prepara- tion, planting density, species selection, competition control, thinning, and other activities. To that end, the lagging public investment in capacity building and technical training must be reversed (Cao, 2008; Yin et al., 2005). Regarding the latter, it is unrealistic to rely on central government investments alone in the long run; local governmental and private entities must play their role throughout the process of ecological restoration (Xu et al., 2006). Of course, as discussed later, this can happen only if the governance is effective and incentives become attractive. In addition, it is essential to adopt an ecosystem management perspective, with emphasis on sustainability, system function, and integrity (Millennium Ecosystem Assessment, 2003) and to better integrate conservation needs into development policies (Loucks et al., 2001).

1.4.2 Strengthening the Governance of Program Implementation

The SFA has been charged with the responsibility of administering the NFPP, the SLCP, and other EREs. At the same time, other agencies responsible for agriculture and livestock production, water and soil conservation, poverty alleviation, and envi- ronmental protection have not been actively involved (Yin et al., 2005). Inter-agency cooperation and coordinated implementation is weak at all levels of the government. Also, the central government has not granted regional authorities and local com- munities more flexibility in implementing the restoration efforts (Bennett, 2008). One may wonder whether this type of top-down, campaign-style push for ecologi- cal restoration, or any other cause, can be truly effective and long lasting, even with the huge financial commitments.

Also, the SFA has assumed the dual roles of program implementation and monitoring. Without a clear separation of monitoring and implementation and the independence of monitoring, effective implementation of the EREs is questionable, and the efficiency of public investment may be compromised (Xiu & Cao, 2001). Given the political structure in China and the bureaucratic tendency of reporting, or even exaggerating, good news while concealing, or disregarding, bad news, this is especially a worrisome situation. To promote effective implementation, thus, an integrated and authoritative program monitoring system must be established, with its own budget and staff. The protocols by which samples are selected, data are collected, and statistics are compiled and released must be scientifically sound. Likewise, transparency and openness of any monitoring effort must be upheld (Xu, 2007).

Currently, monitoring of the socioeconomic and environmental impacts of the EREs is fragmented and incomplete. The SFA Forest Economics and Development Research Center is in charge of monitoring of the socioeconomic impacts of all the EREs; another SFA center assumes the duty of monitoring the desertification and sandification trends; monitoring the short-term changes in land use and resource condition in the program areas is the responsibility of the SFA regional forest inventory apparatuses; and monitoring the long-term ecological conditions and services is largely carried out by the Chinese Academy of Sciences and the Chinese Academy of Forestry, in conjunction with regional universities and research institutes. Lacks of independence, transparency, and adequate procedures are major hurdles to those efforts affiliated with the SFA, whereas lacks of coordination, collaboration, and funding are impediments facing the science community (Xu, 2007). Also, lacks of a scientific advisory and stakeholder representation are common problems of all the monitoring activities. Obviously, systematic assessment of the program effectiveness and impacts becomes difficult given these monitoring problems.

1.4.3 Better Planning and Management

Deficiency of adequate planning, targets in excess of the feasible capabilities of local entities, rush for implementation, and failure to follow through are just some of the issues that hinder real progress (Bennett, 2008; Liu, 2002). More specifically, the drastic logging bans and harvest reductions and thus high employee layoffs under the NFPP could have been gradually executed, which would have alleviated a lot of the losses and traumas inflicted on local economies and stakeholders (Yin et al., 2005). In fact, a number of challenges have arisen from the logging bans and harvest reductions. A primary example is the fact that, in addition to the national forests covered under it, the NFPP also has spread over to community forests in several provinces, including Sichuan, Guizhou, and Yunnan (Yu et al., 2002). In some cases, not only (secondary) natural forests but also plantation forests have been put under the purview of the NFPP. Because of its orientation toward national forests and state forestry enterprises, however, the regulation of collective forest activities and the compensation to community entities were not initially conceived. Accordingly, while farmers' logging and management operations were denied or disrupted, their losses and expenditures on forest protection and management were not subsidized (Yin et al., 2005). These caused a large uproar and only much later were logging restrictions relaxed and protection and management partially incorporated into the government funding. Still, farmers are perplexed by why their commercial activities cannot be allowed and will not benefit the environment as well.

Moreover, few of the other potential consequences of the logging bans and harvest reductions have been carefully considered (Xu et al., 2006). Included in them are the introduction of invasive species, the environmental damage to exporting countries of increased timber imports, and the social costs to rural communities that are adjacent to and thus dependent on the forests covered by the NFPP for seasonal jobs and service opportunities.

For the SLCP, neither the 2002–2004 fervent expansion nor the severe contractions thereafter are conducive to an effective implementation. The government should have foreseen the induced burdens of program coordination, activity inspection, and compensation delivery and thus taken them into account (Xu, Tao, & Xu, 2004). Poor administrative budgeting and capacity caused many of the millions of plots participating in the SLCP yet to be inspected (Bennett, 2008). It is only since 2002 that the central government has allocated any administrative fees to provincial agencies for implementing the SLCP. These, however, have been insufficient and are often in large part diverted by higher levels of government before reaching the local constituents (Bennett, 2008).

Of course, this is partially a result of the fast expansion of the program, which created even greater administrative needs, and thus shortfalls in administrative funds and staff, which in turn led to problems in implementation and subsidy delivery. Consequently, the supervision was weakened, the delivery of food subsidies to farmers was delayed and even deducted, and the quality of project tasks deteriorated (Xu & Cao, 2001). Clearly, in addition to careful upfront planning and piloting, it is necessary to make timely and careful adjustments to the specific targets and measures according to the changed circumstances in order to better implement the programs (Wang, Lu, et al., 2007). As part of the DCBT, some farming and herding families are resettled to places with less degraded land with greater carrying capacity. The question remains, though, how to guarantee the farmland and grassland around the newly resettled sites will not be degraded in the future. Even if the initial response is positive, it is unclear whether the current arrangements will work in the long term.

1.4.4 Emphasis on Local People's Active Engagement

Without adequate consultation and community-based initiatives, local people tend not to plant or to maintain the trees and grass properly. The resultant survival and growth rates can be meager. More broadly, the central authorities have failed to realize the importance of the incentive structure (FAO, 2004). They place too much reliance on administrative campaigns, and not enough on contracts, open bidding, and other market-based mechanisms for carrying out specific activities (Yin et al., 2005). Coupled with the uniform standards of subsidies, this is worrisome because of the potential for compromising effectiveness, sacrificing efficiency, and even aiding misappropriation (Xu et al., 2004). Also, without clearly set responsibilities and appropriate rewards, forest enterprises, state employees, and rural households will not as actively participate in forest, grassland, and wetland protection and management as might be assumed (Du, 2001; Agrawal et al., 2008). Along with this is the need to enhance the adaptive capability and capacity of farmers and herders, so that they can move out of the dependency on a traditional forest or agricultural economy and government subsidies.

Furthermore, private participation and investment must be fully appreciated and actively sought. Without strong and lasting private engagement, sustainable forestry and ecosystem management will not be accomplished, no matter what and how much the government does. China's experience has proven that farmers are interested in and can contribute to an array of forestry and natural resource management activities as long as the incentive structure is arranged properly (Yin et al., 2003). The slow progress of the ITPP only indicates that the incentive structure is such that the forest products companies and rural farmers remain unwilling to engage in commercial forestry (Xu et al., 2006). Naturally, one may wonder why the government can make great financial and programmatic commitments to ecological restoration but cannot have effective policy and institutional adjustments to the incentive structure to attract private interest and action. Indeed, a key component of the incentive structure is not only the definition but also the realization of property rights. While China has made breakthroughs in the former, it still has a long way to go in the latter (Wang, Innes, et al., 2007).

Therefore, the current forest tenure reform in the southern community forest region and the state forest enterprise reform in the northeast should be integrated with the ecological restoration efforts and implemented jointly. Related to this, the notion that the commercial forestry and other activities of private interest will not benefit the environmental protection must be refuted (Yin et al., 2003). The fact is that improved resource condition and productivity attributable to the private sector can both alleviate the government burdens in providing timber, fuel, fodder, and other products, and generate a whole host of ecosystem services like erosion control, watershed regulation, and carbon storage (FAO, 2004).

1.5 Concluding Remarks

In this paper, we have provided a complete description of China's EREs and updated their implementation. It can be seen that China has not only made unprecedented commitments but also impressive progress in implementing some of the world largest ecological restoration programs. We also have deliberated several critical issues that the government must address in order to implement these EREs more effectively and to improve the country's ecosystem functions significantly.

We argue that it is crucial for China to embrace a more balanced and adequate approach to ecological restoration; adopt better planning and management practices; strengthen the governance of program implementation; emphasize local people's active engagement; establish an independent, competent monitoring network; and conduct timely, high-quality assessments of the program effectiveness and impacts. And it is our view that in each of these areas, well-organized and effective scientific research and policy change must be pursued and the international community can and should provide a wide range of assistance.

Finally, it should be reiterated that considering limited space and empirical evidence, we could not have gathered sufficient findings of assessed program impacts in this paper. And to a large extent, our narrative of the program implementation was based on government statistics. As more independent monitoring reports and scientific studies appear in the literature, it will become feasible to review and synthesize them and thus to depict a comprehensive picture of the effectiveness and impacts of China's EREs. This endeavor will benefit China and many other countries in their continued pursuant of ecological restoration and sustainable development.

Acknowledgments This study was funded by the U.S. National Science Foundation (project 0624018). The authors are grateful for the comments and suggestions made by many participants of the International Symposium on Evaluating China's Ecological Restoration Programs, held on October 19, 2007, in Beijing. They appreciate Lanying Li and Erin Shi for their assistance.

References

Agrawal, A., Chhatre, A., & Hardin, R. (2008). Changing governance of the world's forests. *Science, 320*, 1460–1462

Bennett, M. T. (2008). China's sloping land conversion program: Institutional innovation or business as usual? *Ecological Economics, 65*, 700–712

Cao, S. X. (2008). Why large-scale afforestation efforts in China have failed to solve the desertification problem. *Environmental Science and Technology, 42*(6), 1826–1831

Du, S. F. (2001). *Environmental economics.* Beijing: Encyclopedia Press

FAO. (2004). *What does it take? The role of incentives in forest plantation development in Asia and the Pacific.* Bangkok: FAO Regional Office for Asia and the Pacific

Forest and Grassland Taskforce of China (1/2003). *In Pursuit of a Sustainable Green West* (Newsletter)

Liu, C. (2002). *An economic and environmental evaluation of the Natural Forest Protection Program* (working paper). Beijing: The SFA Center for Forest Economic Development and Research

Liu, J. G., & Diamond, J. (2005). China's environment in a globalizing world. *Nature, 435*, 1179–1186

Liu, J. G., Li, S. X., Ouyang, Z. Y., Tam, C., & Chen, X. D. (2008). Ecological and socioeconomic effects of China's policies for ecosystem services. *PNAS, 105*, 9477–9482

Loucks, C. J., Lu, Z., Dinerstein, E., Wang, H., Olson, D. M., Zhu, C. Q., & Wang, D. J. (2001). Giant pandas in a changing landscape. *Science, 294*, 1465

Lu, W. M., Mills, N. L., Liu, J. L., Xu, J. T., & Liu, C. (2002). Getting the private sector to work for the public good: Instruments for sustainable private sector forestry in China. *Instruments for sustainable private sector forestry series.* London: International Institute for Environment and Development

Millennium Ecosystem Assessment (2003). *Ecosystems and human well-being: A framework for assessment.* Washington, DC: Island Press

Normile, D. (2007). Getting at the roots of killer dust storms. *Science, 317*, 314–316

Smil, V. (1993). Afforestation in China. In A. Mather (Ed.), *Afforestation: Policies, planning and progress.* London, UK: Belhaven Press

State Forestry Administration (SFA) (2000–2008). *China forestry development report.* Beijing: China Forestry Press

Tang, X. L. (2007, October 19). *China's ecological restoration programs and policy* Paper presented at the International Symposium on Evaluating China's Ecological Restoration Programs, Beijing

Uchida, E., Xu, J., Xu, Z., & Rozelle, S. (2007). Are the poor benefiting from China's land conservation program? *Environment and Development Economics, 12*(4), 593–620

Wang, C., Ouyang, H., Maclren, V., Yin, Y., Shao, B., Boland, A., & Tian, Y. (2007). Evaluation of the economic and environmental impact of converting cropland to forest: A case study in Dunhua county, China. *Journal of Environmental Management, 85*(3), 746–756

Wang, G. Y., Innes, J. L., Lei, J. F., Dai, S. Y., & Wu, S. W. (2007). China's forestry reforms. *Science, 318*, 1556–1557

Wang, X. H., Lu, C. H., Fang, J. F., & Shen, Y. C. (2007) Implications for development of Grain-for-Green policy based on cropland suitability evaluation in desertification-affected north China. *Land Use Policy, 24*, 417–424

Wei, H. D., Xu, X. Y., Ding, F., & Gao, Z. H. (2006). Dynamic desertification monitoring of minqin oasis over the past five years. In L. K. Zhu (Eds.), *Dynamics of Desertification and Sandification in China* (pp. 193–204). Beijing: China Agriculture Press

WWF (World Wide Fund for Nature). (2003). *Report Suggests China, Grain-to-Green plan is fundamental to managing water and soil erosion.* http://www.wwfchina.org/english/local.php? loca=159

Xu, J. T. (2007). *A preliminary review of the monitoring system of China's major ecological restoration programs* (working paper). Peking: Peking University College of Environmental Science and Engineering

Xu, J. T., & Cao, Y. (2001). Converting steep cropland to forest and grassland: efficiency and prospects of sustainability. *International Economic Review, 2*, 56–60 (in Chinese)

Xu, J. T., Tao, R., & Xu, Z. G. (2004). Sloping land conversion: Cost-effectiveness, structural adjustment, and economic sustainability. *China Economics Quarterly, 4*(1), 139–162 (in Chinese)

Xu, J. T., Yin, R. S., Li, Z., & Liu, C. (2006). China's ecological rehabilitation: Unprecedented efforts, dramatic impacts, and requisite policies. *Ecological Economics, 57*, 595–607

Yin, R. S. (2007, October 19). *An integrative evaluation of China's ecological restoration programs.* Paper presented at the International Symposium on Evaluating China's Ecological Restoration Programs, Beijing

Yin, R. S. (1994). China's rural forestry since 1949. *Journal of World Forest Resource Management, 7*, 73–100

Yin, R. S. (1998). Forestry and the environment in China: The current situation and strategic choice. *World Development, 26*(12), 2153–2167

Yin, R. S., Xu, J. T., & Li, Z. (2003). Building institutions for markets: Experience and lessons from China's rural forest sector. *Environment, Development, and Sustainability, 5*, 333–351

Yin, R. S., Xu, J. T., Li, Z., & Liu, C. (2005). China's ecological rehabilitation: The unprecedented efforts and dramatic impacts of reforestation and slope protection in Western China. *China Environment Series, 6*, 17–32

Yu, Y., Xie, C., Li, C. G., & Chen, B. L. (2002). *The NFPP and its impact on collective forests and community livelihoods.* A report commissioned by the Forest and Grassland Taskforce of China, Beijing

Zan, G. S., & Wang, J. H. (2006). Land cover and sandification tendency of Bashang (Northern Hebei) and the causes. In L. K. Zhu (Ed.), *Dynamics of desertification and sandification in China* (pp. 143–153). Beijing: China Agriculture Press

Chapter 2
Assessing China's Ecological Restoration Programs: What's Been Done and What Remains to Be Done?

Runsheng Yin, Guiping Yin, and Lanying Li

Abstract This article surveys the recent literature that assesses China's ecological restoration programs, including the Sloping Land Conversion Program (SLCP) and the Natural Forest Protection Program (NFPP). Our presumption is that the performance of these programs should be determined by their effectiveness of implementation and significance of impact. Implementation effectiveness can be measured with such indicators as land area converted or conserved, and survival and stocking rates of restored vegetation, while impact significance can be gauged by the induced changes in ecosystem productivity and stability, and socioeconomic conditions. Coupling this matrix with an exhaustive search of the publications, we find that: (1) the implementation effectiveness has not been examined as extensively as the impact significance; (2) efforts to assess the impact significance have concentrated on the SLCP, particularly its socioeconomic effects: growth of income, alternative industry, and employment, and likelihood of re-conversion; and (3) most of the socioeconomic studies are based on rural household surveys, and discrete choice and difference in differences models. Future work should pay more attention to the NFPP and other programs, and the environmental impacts and the implementation effectiveness of all of them. To these ends, analysts must gather more field data regarding the evolving ecosystem conditions and socioeconomic information of higher aggregation, and conduct their research across scales and disciplines, with better application of geospatial technology and more effective collaboration.

Keywords Ecological restoration · Effectiveness of implementation · Environmental and socioeconomic changes · Significance of impact

R. Yin (✉)
Department of Forestry, Michigan State University, East Lansing, MI 48824, USA;
Ecosystem Policy Institute of China, Chaoyang District, Beijing 100102, P.R. China

R. Yin (ed.), *An Integrated Assessment of China's Ecological Restoration Programs*,
DOI 10.1007/978-90-481-2655-2_2, © Springer Science+Business Media B.V. 2009

2.1 Introduction

To alleviate its enormous environmental challenges, to reduce the poverty and improve the livelihoods of its vast rural population, and to adjust its economy structure, China launched some major programs of ecological restoration and resource expansion in the late 1990s and the early 2000s (Yin et al., 2005; Bennett et al., 2004). They include the Sloping Land Conversion Program (SLCP), the Natural Forest Protection Program (NFPP), and the Desertification Combating Program around Beijing and Tianjing (DCBT). The specific contents of these programs, and the accomplishments and challenges of their implementation have been covered elsewhere (Chapter 1). The purposes of this chapter are to provide an up-to-date survey of the recent literature that has assessed these restoration programs and to lay out some essential directions and tasks for integrated assessment of these programs in the future.

Before proceeding, it is especially constructive for us to outline the basic elements upon which these programs are examined and the primary indicators against which they can be assessed. We hope that with such an outline, our review can be better organized and thus straightforward to follow. In our view, the most relevant elements of the performance of these programs are their efficacy of implementation and significance of impact. Implementation efficacy refers to what a program has achieved according to its operational targets, whereas impact significance concerns how a program's execution has served its ultimate missions. The former can be gauged with such indicators as land area converted or conserved, effectiveness of site selection for conversion and preparation for tree/grass planting, and survival and stocking rates of vegetation planted or rehabilitated. The latter can be measured by the induced environmental and socioeconomic changes. Environmental changes are reflected in ecosystem productivity and stability, such as the status of biodiversity, soil erosion, and carbon sequestration. Socioeconomic changes are represented by such indicators as income enhancement, labor transfer, and cost efficiency. Further, these environmental and socioeconomic changes can be classified into short- and long-term effects.

Obviously, there can be alternative ways to characterize the elements and indicators of program efficacy and impact, and our outline above may not constitute an exhaustive list or most appropriate categorization of them. But these considerations are beyond the scope of this chapter. Moreover, as will be seen below, they are the elements and indicators that have attracted the mostly research attention and indeed been relatively more scrutinized.

Notably, several international projects have been conducted to assess the implementation and impact of China's ecological restoration programs. The first one was carried out by the Western China Task Force on Forests and Grasslands (TFFG) of the China Council for International Cooperation on Environment and Development (CCICED)—an advisory body to the central government on environmental affairs. With funding from the World Bank, the CCICED commissioned the TFFG in 2000 to investigate the early performance of the programs. As TFFG's secretary, Jintao Xu from the Center for Chinese Agricultural Policy had the privilege to initiate a series of rural household surveys in several western provinces to spearhead the TFFG charges. Based on these surveys, Xu and his associates have published several empirical studies (e.g., Xu, Katsigris, & White, 2002; Xu et al., 2004; Uchida, Xu, & Rozelle, 2005; Bennett, 2007; Uchida, Xu, Xu, & Rozelle, 2007).

Another significant project was later undertaken by the Forest Economics and Development Research Center (FEDRC) of China's State Forestry Administration and Australian National University (ANU), with financial support from the Australian Center for International Agricultural Research. A bulk of their work was summarized in the project report (Zhang, Bennett, Wang, Xian, & Zhao, 2006) and a book (Bennett et al., 2008), while the researchers have produced some journal articles as well (Xie et al., 2006; Wang, Bennett, Xie, Zhang, & Liang, 2007).

The third project was carried out by the CCICED Task Force on Environmental & Natural Resources Pricing and Taxation, with funding from the British Department for International Development and coordination by Shiqiu Zhang of Peking University. Based on household survey data, Zhang and her collaborators from the United Kingdom also were able to generate a set of empirical analyses, which were reported in Zhang, Swanson, and Kontoleon (2005) and published elsewhere.

Yet another project was done by a team of scientists from Canada and China, with support from the Canadian International Development Agency and the Chinese Academy of Sciences. As part of their assessment of the carbon sequestration potentials in China's forest ecosystems, they examined the SLCP impacts, and their work was highlighted in a recent special issue of *Journal of Environmental Management* (Chen et al., 2007).

Below, we will review these and other works in detail by coupling them with the matrix of assessment we have outlined above. To summarize here, we have found that: (1) a number of studies of these programs have appeared in the international literature,[1] including *Science* and other primary journals; (2) surprisingly, the implementation efficacy has not been examined as extensively as the impact significance; (3) efforts to assess the impact significance have concentrated on the socioeconomic effects, mainly income change, growth of alternative industry and employment, and likelihood of re-conversion; and (4) most of the socioeconomic studies are directed at the SLCP and based on rural household surveys and discrete choice and difference in differences models. Therefore, future work should pay more attention to the environmental impacts and the efficacy of the program implementation. To these ends, analysts must gather more field data regarding the evolving ecosystem conditions and socioeconomic information of higher aggregation, and conduct their work across scales and disciplines and in a spatially explicit manner. Also, more efforts should be directed to assessing the NFPP, the DCBT, and other programs. Of course, the government should provide not only increased funding for but also better coordination of these endeavors.

---

[1] There have been many studies published in domestic Chinese journals. But because most of them are not peer reviewed and less accessible, only a few are covered here.

Because the SLCP and the DCBT overlap geographically and are similar in many other ways (Chapter 1) and because the latter is less scrutinized, they will be combined in our review. Also, it seems more convenient to present the assessed environmental and socioeconomic impacts of the programs separately. In the following four sections, thus, we will examine the implementation efficacy of the programs, the socioeconomic impacts of the SLCP, the socioeconomic impacts of the NFPP, and the environmental impacts of these programs, respectively. Finally, we will articulate some directions and tasks for integrated assessment of the programs in the future.

2.2 Implementation Efficacy

Xu and Cao (2002), Xu, Bennett, Tao, and Xu (2004) and Xu, Tao, and Xu (2004) were among the first describing TFFG's preliminary findings of the efficacy of the SLCP and farmers' initial response to it. Based on case studies conducted in seven provinces, they demonstrated that more than 80% of the cropland plots selected for conversion was on slopes steeper than 15° and that their yields in 1999 were lower compared to those plots that were not retired. This finding was later corroborated by Zhang et al. (2005), which showed that the likelihood of a plot being selected into the SLCP increases as its productivity decreases. Likewise, Uchida et al. (2005) [2] found that 83% of the cropland set aside under the SLCP was on slopes steeper than 15°. Xu and Cao (2002) and Xu, Tao, et al. (2004) also reported a rate of farmers' satisfaction greater than 90% in five of the seven cases, due largely to the food subsidies greater than the average annual household production.[3] They further indicated that because of the heavy subsidies, local government officials had urged farmers to retire much more than what had been planned by the central government.

Nevertheless, these studies asserted that the SLCP could be implemented more effectively. For example, in Dingxi, Gansu, 17% of the cropland set aside under the SLCP was not on slopes of greater than 15°. Further, they noted that some regions gave priority to sites close to roads for conversion, in order to facilitate the inspection and monitoring by upper-level government officials; and in certain cases county governments sought to include plots that are contiguous but not on steep slopes to minimize implemented cost. It is disturbing and even deplorable if sites are indeed selected close to main roads for the purpose of demonstrating accomplishments by the local governments. On the other hand, since the household cropland plots are generally tiny and dispersed, the local decision to implement restoration activities on a larger, landscape scale, even if some flatter plots are included, may be jus-

---

[2]This latter study and others conducted by Emi Uchida were based on Jintao Xu's household survey data and often have him as a coauthor.

[3]Of course, household grain production will entail various inputs and thus expenses, whereas government subsidies are completely free.

tified in view of the integrity of ecosystem functions and the potential for saving monitoring and management costs.

Using Dunhua county in Jilin as a case, Wang, Ouyang, et al. (2007) drew the conclusion that while the local and regional biophysical and socioeconomic characteristics vary greatly in terms of their productivity and susceptibility to soil erosion, this heterogeneity does not appear to have been sufficiently taken into account in site selection. The efforts to expand the SLCP quickly nationwide sacrificed careful planning and failed to assess properly some of the lessons from the pilot projects. Also, they stressed that more attention should be paid to the quality of the forestation rather than its quantity. Therefore, they suggested that officials should consider offering enhanced information and training services and increased public participation in decision making, so that local acceptance and participation can be enhanced. These points were echoed by Bennett (2007).

The NFPP mandated that commercial logging would be completely banned in the upper reaches of the Yangtze and Yellow Rivers by the year 2000 in order to conserve over 61.08 million ha of forest. Timber removals from these basins would decrease from 87.58 million m^3 in 1997 to 26.50 million m^3 in 2000, with most of the remaining harvests to accommodate local fuel, housing, and other noncommercial uses. This would mean a loss of 12.4 million m^3 of annual production of roundwood in these two river basins. Commercial logging in the northeast and elsewhere would be greatly curtailed in order to put 33 million ha of predominantly old-growth forests under protection. Roundwood production in these regions would decrease from 18.54 million m^3 in 1997 to 11.02 million m^3 in 2003. According to the monitoring reports published by the State Forestry Administration (SFA 2002–2006), these goals were generally accomplished.

Zhang et al. (2000) overviewed the logging restrictions induced by the NFPP and their potential environmental as well as economic impacts. They indicated that the investment of the central government was a success in terms of timber production control, land conversion, and resettlement of forest dwellers based on statistics from the first 2-year implementation. Debating the appropriate policy measures for its implementation with Zhang et al. (2000), Xu et al. (2000) pointed out that there exist some flaws with the program, although the NFPP may temporarily mitigate ecological degradation in the areas covered. They argued that the NFPP ignores the scientific method—ecosystem management that emphasizes the sustainability, ecosystem functionality and integrity, and human interaction with the ecosystem. Loucks et al. (2001) asserted that the NFPP could strengthen the conservation of the panda in China's forests by enhancing protection and restoration of corridors among remaining forest fragments and increasing habitat preservation. They also claimed that the NFPP could provide a great opportunity to move panda conservation from individual reserves to habitat conservation across landscape.

Zhao and Shao (2002) discussed the potential environmental effects of the logging restrictions. They insisted that China's logging restriction, the core of the NFPP, will protect millions of hectares of forests from further clearing. In addition, they held that this program can help China adjust the age structure of national forests and provides a new chance for restoring forest sustainability.

How about the survival, stocking, and growth rates of the planted or regenerated trees and grass under different programs? It seems that analysts have deferred to the government agencies for answering this question. Again, the SFA monitoring reports portray a rosy picture: the survival rate of the tree planting efforts was at least 85%, the forest coverage has been increasing steadily, and the tree and grass stocking level has improved markedly (SFA, 2007). However, it should be said that these reports were based on information provided by the local government agencies. Without independent validation and assessment, their accuracy and reliability may be questionable.

2.3 SLCP's Socioeconomic Impacts

Cost effectiveness of the SLCP is an important concern of previous studies. As defined by Uchida et al. (2005), cost effectiveness refers to whether the program is achieving its environmental goals at the lowest possible cost. To evaluate the cost effectiveness of a land set-aside program, it would be ideal to have information from participating and non-participating households and to compare the environmental benefits and opportunity costs of plots from both sets of households. But Uchida et al. (2005) had only information on program participants. As such, they estimated that 40% of the plots enrolled in the SLCP had a yield that was lower than the level of compensation, which implies that there was a significant degree of over-compensation. Based on the purchase power parity, these authors also identified that the average compensation is 50% higher than the budgetary outlay of the Conservation Reserve Program (CRP) in the U.S. These results imply that the government might be able to generate fiscal savings if the payments could be made to more accurately reflect the variation in the opportunity costs of plots. As such, Uchida et al. (2005) and Feng et al. (2005) suggested that the cost effectiveness would improve if the current system that only differentiates the grain payment on a broad regional basis—one scheme for the Yellow River basin and another for the Yangtze River basin—were replaced by payments made according to the opportunity costs of local plots.

In contrast, Xu, Tao, et al. (2004) pointed out that while some households were over-compensated, others were under-compensated. This indicates that the differences between the opportunity costs and the compensatory benefits vary across regions. In addition to reminding readers of the fact that part of the compensation was made explicitly by the central government to partially reduce rural poverty, these authors argued that one of the main reasons for the central government to adopt simplified and more uniform compensation schemes is to counter the rent-seeking behavior of local officials, which might inflate the subsidies by exaggerating opportunity cost estimates.

Several studies have further assessed the impact of the SLCP on the livelihood of local communities. Uchida et al. (2005) found that the average household net income for the SLCP participants has increased significantly, ranging from 75%

in Ningxia to 8% in Guizhou. Using separate survey data, Xie et al. (2006) also found that the SLCP had a positive impact on recipients' revenue. However, Xu, Bennett, et al. (2004), Xu, Tao, et al. (2004) found that over the period of 1999–2003, the growth rates in average net income vary greatly across the surveyed regions of their sample. Incomes of participants and non-participants exhibited a very similar growth rate in Shaanxi, and participant incomes showed a slower increase than that of non-participants in Gansu. But in Sichuan, participant incomes grew more rapidly, compared to their non-participating counterparts. Overall, they showed that the impact of the SLCP on participants' income is statistically insignificant.

Zhang et al. (2005) pointed out that the above studies did not compare the revenue of participants with what it would have been without program participation. A possible reason is that the data sets used contain information from participants only; hence, they did not have data to derive a counterfactual. While the dataset used by Xu, Tao, et al. (2004) had information from non-participants, they assumed that participation in the SLCP is exogenous and thus not influenced by farmers' self choice. Therefore, it is possible that the estimated SLCP impacts on income levels suffered from self-selection bias. So, Zhang et al. (2005) reassessed the impact of the SLCP on recipients' income using both parametric and non-parametric techniques. Interestingly, their results showed that the impact of the SLCP on total income is not significant even at the 10% level.

Using panel data and a fixed-effect model, Liu and Zhang (2006) found a positive impact of converting farmland to forestland on household income in the vicinity of Beijing and Tianjin, where the DCBT has been implemented. According to their estimation, household income would indeed be 17.4% higher if the sample village had introduced the program 1 year earlier. To evaluate the impact of the SLCP and the DCBT on poverty alleviation, Liu and Zhang (2006) also analyzed the relationship between participation and poverty status. They showed that the enrollment into these programs is negatively related to the poverty rates, implying that poverty reduction is not seriously considered in implementing the program.

Zhang et al. (2005) also analyzed the impact of the SLCP on poverty alleviation. Using novel quartile regressions (income in the 25%, 50%, and 75% quartiles), they found that the SLCP is making a significantly positive impact on the incomes of poor farmers. Using different versions of the propensity score method and survey data of 360 households for 1999 and 2003, however, Uchida et al. (2007) found only a moderate success of the SLCP in achieving its poverty alleviation goals. Also, they did not come up with strong evidence in supporting the claim that participating households have begun to shift their efforts into the off-farm wage earning or self-employed activities. Based on the example of Liping county in Guizhou, Zhou et al. (2007) detected the importance of government subsidies as well. They reasoned that because the net revenue generated from timber plantations is a lot less than that from producing annual crops, the provision of subsidies by the government has not only made the project economically feasible, but has also become a major means of elevating farmers' income; in the long run, the tree plantations for most species will generate sufficient economic return given stable market prices, so that the removal of the subsidies will not alter the financial situation of the farmers in the future. In another study, Xu, Yin, and Zhou (2007) also found that the SLCP has contributed to the social transformation of the traditional rural society by enabling the freed-up workers from farming fields to seek off-farm jobs in or outside of the locale.

In addition, scholars have examined the issue of whether participation in the SLCP is voluntary. Zhang et al. (2005) claimed that, in principle, participants select themselves into the program. However, others argued that there is evidence showing that the autonomy of the individual household is incomplete, and that participation is decided by the local government (Bennett, 2007). Xu and Cao (2002) reported that enrollment into the SLCP was completely involuntary, whereas Xu, Yin, Li, and Liu (2006) claimed that only 21% of the households surveyed were consulted before the set-aside decision and only a third had a say in how much cropland or which specific plots to be converted. It appears, though, that involuntary participation occurs when the compensation a household received for setting aside its land does not cover the opportunity cost of farming (Uchida et al., 2005).

The village leader survey taken by Zhang et al. (2005) shed more light on this issue. They reported that out of the 40 interviewed, only three village leaders stated that participation in the SLCP was entirely voluntary. In most of the cases participation was based on a combination of self-selection by households and final selection by the local governments. So, they assumed that households select themselves into the SLCP based on their expected returns, rather than that the local government maximizes environmental benefits minus the opportunity costs of lost agricultural production, as did in Uchida et al. (2005). Therefore, they examined the determinants of household selection into the SLCP with a probabilistic model. The estimated results confirmed that households with lower average productivity are selected by local governments, which is a good sign for the cost effectiveness of the SLCP.

Another important aspect of the short-run impact of the SLCP is the effect of the program on food security, which has been a central topic of debate among policy makers and analysts. Using a multi-objective programming model, Feng et al. (2005) simulated the impact of the SLCP on China's grain supply in the upper reaches of the Yangtze and the Yellow. They found that this impact was in the range of 2–3%, suggesting that the SLCP might not have a major impact on China's future grain supply. But the impact in certain local areas could be significant. Later, Xu, Xu, et al. (2006) insisted that the study by Feng et al. (2005) did not take into account the changes in farmer production behavior on the remaining cultivated land, such as response to price changes, and also did not examine effects by crops. By simulating the production and price effects of the program, however, Xu, Xu, et al. (2006) revealed that the SLCP has an even smaller effect on China's grain production and little influence on prices or food imports.

Research has also tackled the long-run impacts and sustainability of the program (Xie et al., 2006; Uchida et al., 2005; Xu et al., 2004; Ye, Chen, & Fan, 2003). Evidence from similar land set-aside programs in other parts of the world suggests that once payments cease, a non-trivial amount of land may return to its pre-program use (Uchida et al., 2005). The likelihood of re-conversion depends on the opportunity cost of the converted land. An interesting study of this issue is presented by Xie et al. (2006). The authors provided output and prices forecasts for the products of the newly adopted annual and perennial crops to simulate the effect of the commercial benefits of converted areas on recipients' net income over the next 20 years. They found that the potential revenues from converted land are quite attractive, leaving little concern about the sustainability of the program after the government subsidies expire. However, it is far from certain whether or not their production and price assumptions will hold up in the long run. In addition, Zhang et al. (2005) argued for the need to compare the evolution of incomes between participants and non-participants, and they reported that the scope for an increase in the opportunity cost of reforested land over cropland appears to be quite limited since only a small portion of reforested land is commercial forests and the revenues they generated are marginal.

By analyzing the impact of the SLCP in the upper Minjiang River Basin, Ye et al. (2003) also raised the question of how to guarantee farmers a basic living after the 8 years of governmental subsidies, given that the program, in general, has been beneficial in terms of environmental restoration. Uchida et al. (2005) examined the sustainability of the SLCP through its potential in generating sufficient income that will continue after the formal program finishes. This change in participant revenue can be generated by the program directly through benefits derived from commercial forests, or indirectly through a change in the mix of income-generating activities. Consistent with what Uchida et al. (2007) found, these results showed that there have been some changes in the source of household income following the SLCP. For instance, from 1995 to 1999, the number of households gaining revenue from off-farm labor and livestock production in Ningxia has increased on average by 3–4% per year, respectively. However, the authors acknowledged that their analysis is only descriptive.

Xu, Tao, et al. (2004) contended that the ultimate success of the SLCP depends on its ability to restructure the production practices of rural households, so that they can raise the opportunity cost of their non-farm labor (e.g., livestock production and off-farm employment). The authors found evidence for a significant increase in livestock activity for program participants. To the contrary, the study conducted by Zhang et al. (2005) suggested that the number of labor days spent on livestock activity has remained roughly stable in the post-SLCP period, while the revenue from livestock activity was actually found to have declined since the start of the SLCP. Zhang et al. (2005) focused on analyzing the impact of the SLCP on off-farm labor allocation in assessing the sustainability of the SLCP. Combining a difference in differences approach and a switching regression model with unobserved sample separation, they identified that the impact of the SLCP on off-farm labor allocation is negative for unconstrained households, while for constrained households it is positive and significant. This indicates that participants in the SLCP have increased the off-farm labor supply of constrained, and presumably poor, households.

To assess the sustainability of the SLCP more effectively, studies have also asked the question of what a household intends to do with its retired land after the program payments stop. This approach has been commonly used in assessing the sustainability of a land set-aside program. Uchida et al. (2005) first developed a multinomial choice model of post-contract land use and evaluated the post-program land use decision through the use of stated household intentions. Their evidence on the opinions of farmers suggests that the government should be concerned because a number of participants may be planning to reconvert parts of their land back into cultivated area. In their sample, 34% of the participants in Guizhou and 29% in Ningxia would shift their land back into crop use if the government were to stop the payment after 5 years. Zhang et al. (2005), however, pointed out that the analysis done by Uchida et al. (2005) is mainly descriptive and only deals with land use decisions. They used a method that links land use and labor allocation choices explicitly, arguing that the joint analysis of post-program land and labor allocation choices is a more suitable approach to studying the behavior of rural households as these choices are most likely correlated. With a bivariate probit model, Zhang et al. (2005) detected that 15.5% of respondents would return some land back to cropland and 32% would return all land. Regarding respondents' labor allocation decisions, 41% of respondents would maintain the same activity and 51% would increase farming. These results lead to a higher tendency toward reverting to on-farm activity after the SLCP. They further showed that land-use right plays an important role in the sustainability of the program if subsidies were to stop and improving rental rights would enhance the sustainability of the program by enabling farmers to seek off-farm activities.

Additionally, Zhang et al. (2005) investigated the preferences of households over alternative post-SLCP policy such as subsidy, possibility of renting land, land tenure security, and percentage of ecological forest. The analysis indicated that the most important attribute affecting the preference over alternative SLCP policies is the assurance that the subsidies will be delivered. This result was confirmed by Xu, Bennett, et al. (2004) and Bennett (2007) who demonstrated that at least in the early years of the program, subsidies have not been fully delivered to their rightful recipients in several regions, causing widespread shortfalls. Zhang et al. (2005) also revealed that the development of land rental rights appears to make a significant impact on the households' preferences.

2.4 NFPP's Socioeconomic Impacts

Notably, only a few attempts have been made to assess the impacts of the NFPP and related policy issues. While they have provided good background information for those concerned with China's logging bans and their effects, they are mostly qualitative rather than quantitative. FAO/APEC (2001) included China in its study of the impacts and effectiveness of logging bans in natural forests of the Pacific Rim nations. Zhao and Shao (2002) claimed that although the benefits of the NFPP will be felt for generations to come, the cost of implementing this program is extremely high. First, the central government no longer collects tax revenue from logging profits as it did from the state-owned forestry bureaus. Moreover, the government has to provide additional financial supports to the forestry workers who were laid off from the forest bureaus and timber-processing industries (Xu, Yin, et al., 2006).

Using case studies in Xinjiang, Sichuan, and Yunnan provinces, Katsigris (2002) assessed not only the impact of the NFPP on local enterprises but also the impacts on the local governments and communities, and farmers surrounding natural forests. The author noted that the logging restrictions have significantly reduced the tax revenues of local governments from logging profits, which have subsequently had to decrease investment in infrastructure, primary education, health care, and other public domains. She also found that the NFPP had a negative impact on the local communities and farmers surrounding the natural forests under protection, since it did not yet provide any compensation to them. In addition, she suggested that the logging bans have restricted the freedom of farmers to use their own forest resources, even plantations, which has resulted in a decrease in farmers' incentive to invest in forestry and threatened the sustainability of collective forests.

Likewise, Liu (2002) examined the initial economic impacts of the NFPP. With data from several state forest enterprises and surrounding rural communities, he found that the logging restrictions have dramatically reduced forest bureau revenues from timber; accordingly, local taxes have suffered a major blow. Because of government investments in forest management and ecotourism development, however, their revenues from other sources have expanded, and employee incomes have actually risen sharply. He also discovered that farmers in the restricted areas, who are not formal employees of the state enterprises, have witnessed a severe decline in their income as a result of lost seasonal work and indirect service jobs.

In addition, several studies looked into the impact of the NFPP on the potential domestic reduction and international trade of timber. It was found that the harvest reductions have enlarged the gap between domestic timber supply and demand, and the imports of forest production have increased significantly (Zhao & Shao, 2002; Xu, Yin, et al., 2006). While the projected wood production of 13.45 million m^3/yr. from plantations has not materialized, China's imports of forest products rose by more than 35% in 2001, and shot up to 94.46 million m^3 in 2002 (Xu, Yin, et al., 2006). These increased imports have drawn concerns over the possible spread of deforestation to the exporting countries and its effect on the livelihoods of their forest dependent communities (Mayer et al., 2005). Related to this concern are the widespread phenomena of illegal logging and illegal timber trade (Laurance, 2008), which has gained broad international attention.

By estimating the timber supply and demand of China, Cohen, Lee, & Vertinsky (2000) revealed that because of the time lags involved between planting and harvesting, even the aggressive afforestation program cannot ease the shortage of high-quality timber through domestic supply in the short run. However, the impact of the NFPP in different regions varies, because commercial logging in the southwest was completely banned, whereas it was only scaled back in the northeast. And the timber production is expected to either stay at the same level or grow only slightly over the next several years (Yin et al., 2005).

Shen et al. (2006) measured the socioeconomic impacts of the NFPP using an input-output model. They found that the NFPP will expand the annual output of the forest sectors by 5.8 billion yuan and the whole economy by 8.9 billion yuan by 2010. Employment will increase by 0.84 million in the forest sectors and by 0.93

million in the whole economy. Associated with the enormous expansion of forest protection and management are potentially huge contributions to mitigating water runoff, soil erosion, flooding, and biodiversity loss. They concluded that the investments and adjustments are worthwhile, if the program is properly implemented. According to them, though, the challenges are to transform loggers into tree planters and forest managers and to ensure that the financial and institutional commitments by the local and national governments will eventually materialize.

Finally, despite the achievements of the NFPP in some aspects, several studies have pointed out that the program still faces some tough problems (Xu, Yin, et al., 2006; Zhao & Shao, 2002; Xu et al., 2000; Wang, Innes, et al., 2007). These problems include the heavy reliance on the financial support of the central government, the insufficient involvement of local people in the program implementation, the neglect of long-term planning, and the inconsistency of other related policy measures.

2.5 The Environmental Impacts

Wang, Bennett, et al. (2007) assessed the non-market value of the environmental changes driven by the SLCP, as reflected in improved landscape, water quality, and species abundance, and reduced dust storms. Their work was based on an indirect approach—stated preferences of individual citizens. They estimated that each household in Beijing, Xi'an, and Ansi would be willing to pay 882.56 yuan, 342.56 yuan, and 388.08 yuan, respectively, for improving the natural environment in the Loess Plateau area. An interesting finding notwithstanding, it did not directly look into the primary problems of water runoff and soil erosion changes, or the status of any other ecosystem services.

While they attempted an integrative assessment of the SCLP, the studies led by the Canadian scholars considered mainly net primary productivity and net ecosystem productivity, and their samples were small (Chen et al., 2007). Similarly, with a focus on assessing the co-benefit of carbon sequrestration, they did not examine flooding control, soil erosion, and other acute problems. Nonetheless, a unique feature of these studies was their use of geospatial technology and satellite images in determining the induced land-use changes.

Another study performed by FEDRC and ANU (Jia et al., 2006) did address the question of water balance induced by the SLCP land retirement and conversion. They first divided the study region along the Yellow River basin into small units and estimate the induced water redistributions over time and space. Then, they aggregated the estimates into a regional outcome. The upshot is that during the period of 2000–2020 the SLCP would lead to a reduction of water runoffs of 450 million m^3, equivalent to 0.76% of the total surface water resources. Even though the findings are preliminary, coarse, and not spatially explicit, they are intriguing and suggest a very small impact on the water balance in this semiarid and arid region where water is scarce. As the authors admitted, however, while their work implies a reduction of soil erosion and sedimentation, they did not quantify this effect.

Using a hydrological model, another study (Sun et al., 2006) estimated the potential magnitude of annual water yield response to forestation in China. It suggests that the average water yield reduction may range from about 50 mm/yr (50%) in the semiarid Loess Plateau region in the north to about 300 mm/yr (30%) in the tropical southern region. If these estimates are accurate, they represent large reductions in the annual water yield and thus indicate a great impact of forestation on water flows, particularly on headwaters in the semiarid region. However, the authors cautioned that because the forestation area is relatively small for most large basins with mixed land uses, the regional effects of forestation on water resource management may not be of major concern. Obviously, more systematic evaluation of the roles of forest and grassland on regulating regional water resources is urgently needed.

Despite little formal investigation, several experts came to a shared view that planting tall trees in the semiarid and arid northwestern regions may not work well for the environment; indeed, some even argue that it is detrimental to the environment in the long run (Long et al., 2006; Normile, 2007; Cao, 2008). Among other things, ecologists maintain that planting poplars, a major species for afforestation, in arid regions does not help combat desertification, because of the limited precipitation. Where poplars become established, the deeply rooted trees hemorrhage water through transpiration, lowering the water table and making it harder for native grass and shrubs to survive. Therefore, some insist that it is time for the government to carefully evaluate this policy of species and site selection.

Cao (2008) further asserted that afforestation projects are causing, instead of mitigating, environmental degradation in the arid and semiarid regions, with increased ecosystem deterioration and wind erosion. Therefore, he called for focusing on restoring natural ecosystems that are more suitable to local environments and thus provide a better chance of combating desertification. Clearly, these are personal opinions; but they raise the fundamental question of how to restore degraded land: Should it be planted with trees or grass, or should it be left to natural regeneration? The fact is that the Chinese State Forestry Administration has been enthusiastic about tree planting and regeneration, but not necessarily so with restoring grass coverage or natural re-vegetation. On the other hand, more careful investigations have shown that most of the dryland areas in the west have an annual rainfall less than 400 mm and thus are only suitable for growing grass and drought-tolerant shrub (Wang, Lu, et al., 2007; Chen et al., 2007).

2.6 Synthesis and Future Research

In summary, a number of attempts have been made to assess the implementation effectiveness and impact significance of China's ecological restoration programs. And from the previous efforts, a great deal has been learned. Overall, it appears that the programs have made great strides in achieving their short-term goals, and they have positively affected the economy, environment, and society (Wang & Bennett, 2008; Wang, Innes, et al., 2007; Xu, Xu, et al., 2006). However, much more efforts, especially well-integrated, systematic ones, are warranted to continue assessing these huge undertakings. Compared to the NFPP and other programs, the SCLP, particularly its potential socioeconomic impacts, has received most of the research attention. While this is understandable given its geographic scale and magnitude of investment, other programs deserve greater attention.

Likewise, more efforts should be directed toward understanding the short- and long-term environmental effects of the programs. Unfortunately, many previous works have shied away from the overarching question: Have the environmental conditions been improved by implementing the programs; and if so, how? Indeed, many of them have not even thoroughly examined how successfully forests and grasslands have been established and protected, and how the land use patterns have been affected. Within the socioeconomic realm, certain elements have not been explored as well. For instance, we do not know how the state logging employees and facilities and surrounding rural communities have been affected by the NFPP harvest restrictions. Finally, scholars from different disciplines—ecology, hydrology, economics, sociology, remote sensing, and the like—should collaborate more effectively in their endeavors of assessment.

Therefore, assessing China's ecological restoration programs can and should be expanded in several directions. First, it is crucial to directly measure the environmental conditions and benefits. To our knowledge, numerous studies on cost-efficiency of the SLCP have used a single biophysical attribute, the land steepness, as a proxy for the environmental condition (Xu, Bennett, et al., 2004; Uchida et al., 2005; Zhang et al., 2005). As environmental benefits and economic costs of ecological restoration vary across the landscape due to differences in climate, topography, land cover, and management practices, we doubt that steepness alone is a sufficient proxy for erosion severity or any other ecological condition. While it is understandable to use such a proxy, particularly during the early years of assessment when ecological and hydrological observations were not available, it should be made clear that this type of indicator has only limited validity and thus direct observations of soil erosion, water runoff, and other biophysical dimensions should be made. As attempts to address this deficiency, the studies by Long et al. (2006) and Wang, Lu, et al. (2007) have developed either a feasibility index or a suitability system for cropland set-aside and re-vegetation, based on both biophysical and socioeconomic conditions.

Some national and regional monitoring networks have been set up now, but there is a long way to go in documenting the ecological conditions, including the establishment of their baselines. More generally, it is greatly desirable to better know the effects of the programs on soil erosion, deforestation, desertification, water quality, and biodiversity preservation. In this context, particularly for regional-level environmental assessment, it immediately becomes necessary to apply geospatial techniques and remote sensing images in quantifying the potential temporal and spatial changes. The advancement of the technology and the cost and resolution of the images are such that it is very feasible to adopt them now.

Second, examining the U.S. CRP experience is beneficial, at least from the perspective of program implementation. Since 1990, the CRP has been using the environmental benefits index (EBI) based on multiple environmental benefits to choose land parcels for enrollment (Cooper & Osborn, 1998). The EBI is a sum of six factors that measure the water quality benefits, air quality benefit, soil erosion, wildlife benefits and other environmental benefits. Studies have shown that the use of EBI increased environmental benefits related to costs. So far, however, there is no EBI or similar procedure applied to the SLCP or the DCBT. In addition to guiding parcel enrollment, this index can be used for monitoring the performance of enrolled parcels as well. Future research ought to explore the adoption of this type of index in China.

Third, studies of the SLCP and the DCBT suggest that there is considerable room for improving the cost-effectiveness by modifying the design and implementation of the subsidy payment. In the current system, the government compensates farmers who enroll in the programs based on a uniform basin-wide rate. Several works have made it clear that flexible payment mechanisms and competitive selection processes (such as auction) would greatly improve the cost effectiveness of the programs. It is also suggested that the SLCP and the DCBT may adopt the bidding process that has been used in the CRP. However, it should be recognized that perfect targeting typically cannot be achieved in practice since transaction costs are involved in collecting and processing information. One main problem that arises from the bidding mechanism for the CRP contracts is the strategic bidding process, which affects the rental rates. Landowners with especially high EBI scores but especially low opportunity costs for their land can potentially submit rent bids above their reservation rents and still have their lands accepted into CRP. In fact, the bidding mechanism may not a realistic option in rural China where the administrative costs to set up such a mechanism would be fairly high. Therefore, the adoption of a more practical payment schedule, such as differentiating compensations based on the benefits of certain plot type, could improve the cost efficiency. So far, little empirical work has been done to evaluate the feasibility and effects of those alternative mechanisms. These gaps should be filled in the future.

Fourth, previous studies emphasized the use of household-level data from surveys. The micro-level data, coupled with discrete choice and difference in differences models, give socioeconomic analysts a great opportunity to examine how local farmers respond to the programs and what the induced consequences are. In fact, much knowledge has been gained from these works. However, it should be pointed out that most of the datasets used in previous studies cover a very short time span and were collected from different, and often few, locations under variable enumerating protocols. In combination with different analytic techniques, these data limitations may have partially caused the divergence, insignificance, or conclusiveness of the research outcomes. As such, accumulating longer time series, expanding sampling coverage, developing common data generation and processing protocols, and refining modeling procedures will be some very promising avenues of continued micro-econometric investigation of the program socioeconomic impacts.

Nonetheless, we should realize that this type of data and the corresponding analysis cannot completely capture land-use and socioeconomic changes at a higher level of aggregation, which are equally interesting and useful, let alone be compatible with assessing the accompanying environmental impacts. Moreover, certain environmental consequences can be examined only on a large geographic scale. Accordingly, aggregate data will be needed to represent the overall response and reflect the spatial heterogeneity. We hope that analyses at the regional or watershed level will soon contribute to an improved understanding of the environmental impacts of the restoration programs.

Fifth, the success of the restoration programs will depend not only on the program stipulations, but also on other related policies and market conditions (Yin et al., 2005; Wang, Innes, et al., 2007). For instance, the price of timber could affect farmers' expectation of their future income and hence change their behavior. Also, the way local farmers manage their livestock, be it open grazing, fenced rotation, or pen feeding, has a major impact on the implementation of the program (Chapter 1). In addition, policies concerning rural land and labor markets can alter the opportunity cost of land use and eventually influence the incentive of farmers to participate in a program. It is thus important to explore the interaction of different policies to properly assess the effectiveness of a program. More work is needed to determine where and how regulations can be improved and how a coherent and consistent incentive structure can be established.

Lastly, a remaining challenge is how to couple the environmental and poverty reduction objectives of the programs. All of the restoration programs are treated as a mechanism to improve environmental quality and provide financial aid to the poor. Achieving an optimal solution to a problem with multiple goals needs to use multiple instruments. This suggests that it may be necessary to investigate the extent to which these goals are compatible with each other. It is still unclear how the poverty alleviation objective has influenced the environmental objective, and whether and how the outcome can be improved by using multiple instruments.

Acknowledgments This study was funded by the U.S. National Science Foundation (project 0624018). The authors are grateful for the comments and suggestions made by four reviewers of this journal and many participants of the International Symposium on Evaluating China's Ecological Restoration Programs, held on October 19, 2007, in Beijing. They appreciate Erin Shi for her assistance.

References

Bennett, J. Wang, X., & Zhang, L. (Eds.). (2008). *Environmental management in China: Land use management*. Cheltenham: Edward Elgar

Bennett, J., Zhang, L., Dai, G. C., Xie, C., Zhao, J. C., Li, D., et al. (2004). *A review of the program for conversion of cropland to forests and grassland.* (Research Report No. 2), Australian Center for International Agricultural Research (ACIAR)

Bennett, M. T. (2007). China's sloping land conversion program: institutional innovation or business as usual? *Ecological Economics, 65*, 700–712

Cao, S. X. (2008). Why large-scale afforestation efforts in China have failed to solve the desertification problem. *Environmental Science and Technology, 42*(6), 1826–1831

Chen, J. M., Thomas, S. C., & Yin, Y. Y. (Eds.). (2007). Carbon sequestration in China's forest ecosystems. *Journal of Environmental Management, 85*(3), 513–628

China Forestry Yearbook (CFY) (2003–2004). Beijing: China Forestry Press (in Chinese)

Cohen, D., Lee, L., & Vertinsky, L. (2000). *China's natural forest protection program (NFPP): Impact on trade policy regarding wood* (working paper)

Cooper, J. C., & Osborn CT (1998). The effect of rental rates on the extension of conservation reserve program contracts. *American Journal of Agricultural Economics, 80*(1), 184–194

FAO/APEC (Asia-Pacific Forestry Commission) (2001). *Forests out of bounds: Impacts and effectiveness of logging bans in natural forests in Asia-Pacific*. Bangkok, Thailand

Feng, Z., Yang, Y., Zhang, Y., Zhang, P., & Li, P. (2005). Grain for Green policy and its impacts on grain supply in west China. *Land Use Policy, 22*, 301–312

Jia, Y. W., Zhou, Z. H., Qiu, Y. Q., Yan, D. H., Zhang, L., Xie, C., et al. (2006). The potential effect of water yield reduction caused by land conversion in the Yellow River basin. In L. Zhang, J. Bennett, X. H. Wang, C. Xie, & J. C. Zhao (Eds.), *A study of sustainable use of land resources in Western China*. Report jointly prepared by the National Forest Economics and Development Research Center, Australian National University, and Australian Center for international Agricultural Research

Katsigris, E. (2002). The local socioeconomic impacts of natural forest protection program. In J. Xu, E. Katsigris, & T. White (Eds.), *The local socioeconomic impacts of natural forest protection program and the sloping land conversion program*. CCICED-WCFGTF, Beijing: China Forestry Press

Laurance, W. E. (2008). The need to cut China's illegal timber imports. *Science, 319*, 1184–1185

Liu, C. (2002). *An economic and environmental evaluation of the Natural Forest Protection Program*. Working Paper, SFA Center for Forest Economic Development and Research

Liu, C., & Zhang, W. (2006). Impacts of conversion of farmland to forestland program on household income: Evidence from a sand control program in the vicinity of Beijing and Tianjin. *China Economics Quarterly, 23*, 273–290 (in Chinese)

Long, H. L., Heilig, G. K., Wang, J., Li, X. B., Luo, M., Wu, X. Q., et al. (2006). Land use and soil erosion in the upper reaches of the Yangtze River: Some socio-economic considerations on China's Grain-for-Green program. *Land Degradation & Development, 17*, 589–603

Loucks, C. J., Lu, Z., Dinerstein, E., Wang, H., Olson, D. M., Zhu, C. Q., et al. (2001). Giant pandas in a Changing landscape. *Science, 294*, 1465

Mayer, A. L., Kauppi, P. E., Angelstam, P. K., Zhang, Y., & Tikka, P. M. (2005). Importing timber, exporting ecological impact. *Science, 308*, 359–360

Normile, D. (2007). Getting at the roots of killer dust storms. *Science, 317*, 314–316

Shen, Y. Q., Liao, X. C., & Yin, R. S. (2006). Measuring the socioeconomic impacts of China's Natural Forest Protection program. *Environment and Development Economics, 11*(6), 769–788

State Forestry Administration (SFA) (2002–2007). *China Forestry Development Reports*. Beijing: China Forestry Press (in Chinese)

Sun, G., Zhou, G. Y., Zhang, Z. Q., Wei, X. H., McNulty, S. G., & Vose, J. M. (2006). Potential water yield reduction due to forestation across China. *Journal of Hydrology, 328*, 548–558

Uchida, E., Xu, J. T., & Rozelle, S. (2005). Grain for Green: Cost-effectiveness and sustainability of China's conservation set-aside program. *Land Economics, 81*(2), 247–264

Uchida, E., Xu, J., Xu, Z., & Rozelle, S. (2007). Are the poor benefiting from China's land conservation program? *Environment and Development Economics, 12*(4), 593–620

Wang, C., Ouyang, H., Maclren, V., Yin, Y., Shao, B., Boland, A., et al. (2007). Evaluation of the economic and environmental impact of converting cropland to forest: A case study in Dunhua county, China. *Journal of Environmental Management, 85*(3), 746–756

Wang, G. Y., Innes, J. L., Lei, J. F., Dai, S. Y., & Wu, S. W. (2007) China's forestry reforms. *Science, 318*, 1556–1557

Wang, X. H., & Bennett, J. (2008). Policy analysis of the conversion of cropland to forest and grassland program in China. *Environmental Economics and Policy Studies, 9*(2), 119–143

Wang, X. H., Bennett, J., Xie, C., Zhang, Z. T., & Liang, D. (2007). Estimating non-market values of the conversion of cropland to forest and grassland program: A choice modeling approach. *Ecological Economics, 63*, 114–125

Wang, X. H., Lu, C. H., Fang, J. F., & Shen, Y. C. (2007) Implications for development of Grain-for-Green policy based on cropland suitability evaluation in desertification-affected north China. *Land Use Policy, 24*, 417–424

Xie, C., Zhao, J. C., Liang, D., Bennett, J., Zhang, L., & Dai, G. C. (2006). Livelihood impacts of the conversion of cropland to forest and grassland program. *Journal of Environmental Planning and Management, 49*(4), 555–570

Xu, J. T., & Cao, Y. Y. (2002). Efficiency and sustainability of converting cropland to forest and grassland in the western region. In J. Xu, E. Katsigris, & T. White (Eds.), *Implementing the natural forest protection program and the sloping land conversion program: Lessons and policy implication* CCICED-WCFGTF (pp. 71–81). Beijing: China Forestry Press. (In Chinese)

Xu, J. T., Katsigris, E., & White, T. A. (Eds.). (2002). *Implementing the natural forest protection program and the sloping land conversion program: Lessons and policy recommendations.* Beijing: China Forestry Press (in Chinese)

Xu, J. T., Tao, R., & Xu, Z. G. (2004). Sloping land conversion: Cost-effectiveness, structural adjustment, and economic sustainability. *China Economics Quarterly, 4*(1), 139–162 (in Chinese)

Xu, J. T., Yin, R. S., Li, Z., & Liu, C. (2006). China's ecological rehabilitation: Unprecedented efforts, dramatic impacts and requisite policies. *Ecological Economics, 57*, 595–607

Xu, M., Qi, Y., Gong, P., Zhao, G., Shao, G. F., Zhang, P. C., et al. (2000). The new forest policy. *Science, 289*, 2049–2050

Xu, W., Yin, Y. Y., & Zhou, S. (2007). Social and economic impact of carbon sequestration and land use change on peasant households in rural China: A case study of Liping, Guizhou Province. *Journal of Environmental Management, 85*(3), 736–745

Xu, Z. G., Bennett, M. T., Tao, R., & Xu, J. T. (2004). China's sloping land conversion program four years on: Current situation and pending issues. *International Forestry Review, 6*(3–4), 317–326

Xu, Z. G., Xu, J. T., Deng, X. Z., Huang, J. K., Uchida, E., & Rozelle, S. (2006). Grain for Green versus grain: Conflict between food security and conservation set-aside in China. *World Development, 34*, 130–148

Ye, Y. Q., Chen, G. J., & Fan, H. (2003). Impacts of the Grain for Green project on rural communities in the upper Min River Basin, Sichuan, China. *Mountain Research and Development, 23*, 345–352

Yin, R. S., Xu, J. T., Li, Z., & Liu, C. (2005). China's ecological rehabilitation: The unprecedented efforts and dramatic impacts of reforestation and slope protection in western China. *China Environment Series, 6*, 17–32

Zhang, L., Bennett, J., Wang, X. H., Xian, C., & Zhao, J. C. (2006). *A study of sustainable use of land resources in Western China.* (Report jointly prepared by the National Forest Economics and Development Research Center, Australian National University, and Australian Center for international Agricultural Research)

Zhang, P. C., Shao, G. F., Zhao, G., Le Master, D. C., Parker, G. R., Dunning, Jr. J. B., et al. (2000). China's forest policy for the 21st century. *Science, 288*, 2135–2136

Zhang, S. Q., Swanson, T., & Kontoleon, A. (2005). Impacts of compensation policies in reforestation programs. Report of the Environment and Poverty Programme to China Council for International Cooperation in Environment and Development

Zhao, G., & Shao, G. F. (2002). Logging restrictions in China: A turning point for forest sustainability. *Journal of Forestry, 100*, 34–37

Zhou, S., Yin, Y., Xu, W., Ji, Z., Caldwell, I., & Ren, J. (2007). The costs and benefits of reforestation in Liping County, Guizhou Province, China. *Journal of environmental management, 85*(3), 22–735

Chapter 3
Methodology for an Integrative Assessment of China's Ecological Restoration Programs

Runsheng Yin, David Rothstein, Jiaguo Qi, and Shuguang Liu

Abstract While research projects have been conducted to examine the impacts and effectiveness of China's ecological restoration programs, few of them represent integrated, systematic efforts. The objective of this chapter is thus to articulate and outline a methodology for an integrative assessment, which, we believe, should embrace both the environmental and socioeconomic changes and engage investigations at multiple scales. Further, these investigations should be pursued through interdisciplinary collaboration with expertise from ecology, economics, hydrology, and geospatial, climate, and land change sciences. We argue that the deployment of geospatial capability, the use of longitudinal data, and the connection between science and policy should be the hallmarks of an integrative assessment. We also describe our general approach and specific models to quantify the environmental and socioeconomic impacts induced by implementing the restoration programs, and address the issue of how to overcome the challenges in generating the data needed for executing various empirical tasks. We hope that the adoption and application of this methodology will make a valuable contribution to a more robust and timely assessment as well as implementation of the ecological restoration programs in and outside of China.

Keywords Integrative assessment · Environmental and socioeconomic impacts · Implementation effectiveness · Treatment effect analysis · Ecosystem service evaluation · Longitudinal data

3.1 Introduction

China has been implementing several major ecological restoration programs since the late 1990s (see Chapter 1). It is of great intellectual interest and broad policy significance to assess the potential environmental and socioeconomic impacts

R. Yin (✉)
Department of Forestry, Michigan State University, East Lansing MI 48824, USA,
Ecosystem Policy Institute of China, Chaoyang District, Beijing 100102, P.R. China

R. Yin (ed.), *An Integrated Assessment of China's Ecological Restoration Programs*,
DOI 10.1007/978-90-481-2655-2_3, © Springer Science+Business Media B.V. 2009

of these programs as well as the effectiveness of their implementation (Forest and Grassland Taskforce, 2003; Xu, Yin, Li, & Liu, 2006; Bennett, Wang, & Zhang, 2008). Doing so will benefit our understating of the performance of these programs and help improve their implementation, as well as contribute to the science of assessing ecological restoration efforts, which is gaining importance and currency given that more and more such efforts are being launched worldwide (FAO, 2006). Certainly, this kind of research is relevant at a time when the international community strives to increase its collective ability to anticipate and manage the complex consequences of global change and to accomplish more sustainable development (IPCC, 2007; Millennium Ecosystem Assessment, 2005).

In fact, Chinese institutions and international organizations have undertaken a number of research projects in this direction. However, as reviewed in Chapter 2, few, if any, of the previous studies represent a truly integrated, systematic assessment of the programs. And most of these projects have not been tied closely with or taken advantage of the recent research advances in treatment effect analysis (Lee, 2005) and ecosystem services evaluation (Liu, Loveland, & Kurtz, 2004). Therefore, a scientifically sound and intrinsically cohesive methodology is urgently needed. The objective of this paper is to articulate and outline such a methodology. We hope that our effort will make a valuable contribution to a more robust and timely assessment as well as a more effective implementation of the ecological restoration programs in China.

It should be made clear at the outset that while we hope what we present in this chapter will be applicable and beneficial to other parts of the world and other type of programs, our focus is on China's ecological restoration programs. Furthermore, we do not even claim to prescribe a "one-size-fits-all" paradigm in the Chinese context. Instead, our intention here is to summarize the methodology that we have adopted in conducting our research project, *An Integrative Impact Assessment of China's Ecological Restoration Programs*, which is funded by the U.S. National Science Foundation. Of course, we have benefitted and will continue to benefit from this methodology. But just like it can be useful to many others engaged in similar scientific endeavors, it can be improved by our peers' critique, which is exactly what has motivated us for its publication.

Also, it should be pointed out that, throughout this paper, our discussion will use the Sloping Land Conversion Program (SLCP) and the Natural Forest Protection Program (NFPP) as examples. We believe that this is reasonable given the fact that they are the two largest programs and that they can well reflect most of the issues of our concern (see Chapter 2). The chapter is organized as follows: We outline a basic framework in the next section; then, we present our empirical approach in Section 3.3 and assessment of environmental and socioeconomic impacts and implementation effectiveness in Sections 3.4 and 3.5; we discuss issues related to data generation in section 3.6; and finally we synthesize our methodology and spell out the actions to be taken in its application.

3.2 A Basic Framework

Broadly speaking, the impacts of ecological restoration programs are manifested in both environmental and socioeconomic changes. Environmental changes are reflected in ecosystem productivity and stability, such as the status of biological diversity, soil erosion, and carbon storage; socioeconomic changes are represented by such indicators as cost effectiveness, labor transfer, and livelihood enhancement. Thus, an integrative assessment must embrace both the environmental and socioeconomic changes. Further, because many socioeconomic and ecological issues are interrelated and crosscut, they must be examined together. For instance, poor soil fertility caused by erosion leads to poor crop productivity, which will in turn drive more extensive farming, resulting in more severe erosion. Figure 3.1 is a sketch of our research framework and tasks.

To assess both environmental and socioeconomic changes induced by carrying out an ecological restoration program, it is essential to develop strong interdisciplinary collaboration with expertise from ecology, economics, hydrology, and geospatial, climate, and land change sciences. Undoubtedly, as much as economists and social scientists lack expertise and experience in addressing ecological and hydrological issues, ecologists and hydrologists may not be well equipped and competent at dealing with social and economic questions. To advance knowledge and solve problems, therefore, effective communication and cooperation of experts from different fields are imperative. Likewise, close collaboration between scientists and practitioners is beneficial (Foley et al., 2005; Boyd, 2007).

Moreover, assessing the impacts and effectiveness of ecological restoration must be conducted at multiple scales. This is because all issues cannot be well examined at a single scale. For instance, some socioeconomic impacts can be easily detected at the household or village level; however, ecological impacts, determined by the integrity of ecosystem functions, must be assessed at least at the watershed scale.

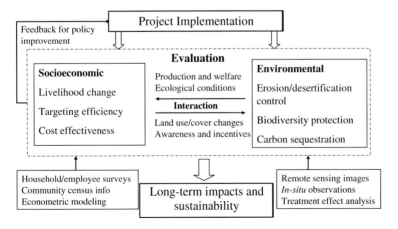

Fig. 3.1 A Summary of the Research Framework and Tasks

Certainly, place-based and regionally focused work is more adequate to accommodate socioeconomic and ecological investigations (Clark, 2007). Indeed, for the major restoration programs in China, a regional-level assessment is more sensible and meaningful, given the personnel, financial, and time constraints involved in a national assessment on the one hand, and the impossibility to answer most relevant questions with local-level investigations on the other. Additionally, research at different scales may lead to different findings that can be complementary (Turner, Lambin, & Reenberg, 2007). As such, a proper way is to have assessment at the multiple scales for multiple tasks.

3.3 The Empirical Approach

Since the 1990s, much progress has been made in treatment effect analysis (TEA) by statisticians, econometricians, and others (Imbens & Woodridge, 2007; Angrist & Krueger, 1999; Heckman, Lalonde, & Smith, 1999). It is often necessary to know the effect of a "treatment" on a response of interest. For example, the treatment can be a drug, an education program, or an economic policy; and the response can be an illness, academic achievement, or gross domestic product. The essence of TEA is to estimate the treatment effect appropriately and adequately, because a response may be due to factors other than the treatment (Lee, 2005). Once the effect is found, one can adjust the treatment to attain the desired level of response.

The TEA approach fits the context of our program assessment well. Here, treatments are the specific activities brought about by such restoration programs as the NFPP and the SLCP; effects are the environmental and socioeconomic impacts we intend to evaluate. To determine whether the SLCP grain and cash subsidies have affected farmers' income from and employment in different activities, for instance, income or employment of the "affected" and "control" groups may be related to their status of participation and other factors before and after the program's introduction.

In general, we wish to evaluate the effect of a *treatment* on an outcome Y over a population of individuals. There are two groups indexed by treatment status $T = 0, 1$ where 0 indicates individuals who do not receive treatment, i.e., the *control group*, and 1 indicates individuals who do receive treatment, i.e., the *treatment group* (Imbens & Woodridge, 2007). We observe individuals in two time periods, $t = 0, 1$ where 0 indicates a time period before the treatment group receives treatment, i.e., *pre-treatment*, and 1 indicates a time period after the treatment group receives treatment, i.e., *post-treatment*. Each observation is indexed by the letter $i = 1, \ldots, N$; individuals have two observations each, one pre-treatment and one post-treatment. Let $\overline{Y}_0^T$ and $\overline{Y}_1^T$ be the sample means of the outcome for the treatment group before and after the treatment, respectively; and likewise, let $\overline{Y}_0^C$ and $\overline{Y}_1^C$ be the sample means of the outcome for the control group before and after the treatment, respectively. Then the outcome Y_i can be modeled by the following equation

$$Y_i = \alpha_0 + \alpha_1 T_i + \alpha_2 t_i + \alpha_3 (T_i \cdot t_i) + \varepsilon_i \qquad (3.1)$$

where α_0, α_1, α_2, α_3 are the coefficients to be estimated and ε_i is the error term. It can be seen that α_1 is the treatment group specific effect (accounting for the average permanent differences between treatment and control), α_2 the time trend common to control and treatment groups, and α_3 the true effect of treatment (Imbens & Woodridge, 2007).

The purpose of our program assessment is thus to find an unbiased estimate of α_3, $\hat{\alpha}_3$, with the data available. Since $\hat{\alpha}_3$ can be interpreted as the difference in average outcome in the treatment groups before and after treatment *minus* the difference in average outcome in the control group before and after treatment, i.e., $\hat{\alpha}_3 = \bar{Y}_1^t - \bar{Y}_0^T - (\bar{Y}_1^C - \bar{Y}_0^C)$, it is also referred to as a "difference in differences (DID)" estimator (Lee, 2005; Imbens & Woodridge, 2007). Whether or not the estimated $\hat{\alpha}_3$ is significant will lead to the acceptance or rejection of the underlying hypothesis.

If either $\hat{\alpha}_1$ or $\hat{\alpha}_2$ is significantly different from 0 but it is not captured by the model, $\hat{\alpha}_3$ will be biased. So, a great deal of attention in program assessment is devoted to estimating $\hat{\alpha}_1$ and $\hat{\alpha}_2$ properly such that $\hat{\alpha}_3$ unbiased. To control for possible omitted variable bias of the *OLS* estimation and to check for the result robustness, the above model can be estimated using the fixed-effect or random-effect technique with panel data. This is particularly important in view of the biophysical and socioeconomic heterogeneity of our study sites (Lee, 2005). In addition, its performance may be affected by the existence of agent self selection, or endogenous choice. In that case, we can consider the choice as a function of other relevant exogenous variables first and then use the "corrected" values of choice to estimate the model (Wooldridge, 2000).

A caveat is that the assumption of a parallel time trend between the treatment group and the control group may not hold, in which case the above DID procedure will no longer be valid (Uchida et al., 2007). As one alternative, the propensity score matching method does not require the parallel trend assumption; that is, it estimates the propensity score of the different groups first and then compares the outcomes of those who have similar propensity scores. This method allows the analyst to match the treatment group and the control group when their observable characteristics are continuous or when the set of explanatory factors that determine the treatment contains many variables (Uchida et al., 2007).

Notably, a few scholars (e.g., Xu, Tao, & Xu, 2004; Zhang, Swanson, & Kontoleon, 2005; Uchida et al., 2007) have employed the DID procedure in their studies of the SLCP impacts on farmers' income, targeting of cropland conversion, and other related issues. While their results are interesting, their analytic scope was narrow and their data series were short. And they, or any other scholars for that matter, have not applied it to assessing the socioeconomic effects of the NFPP, let alone the environmental impacts of the restoration programs. Therefore, it is crucial to bridge these gaps by incorporating all of the elements of interest into a unified TEA framework and testing relevant hypotheses.

The general procedure of our analytic work is to: (1) identify the responses, (2) assess the treatment effects, and (3) make policy inferences. Assessing the treatment effects hinges on the analytic models we use and identifying the responses, including both socioeconomic and environmental responses, has a lot to do with the coverage and quality of the data, it is necessary to discuss our modeling methods and data generation in more detail.

3.4 Assessing Environmental Impacts and Effectiveness

For an environmental change, it is more involved to identify the actual response than to estimate the treatment effect of a program. So, here we will concentrate on describing how to identify the actual response—a specific ecological change occurred during the relevant period of time. To that end, we will use the General Ensemble Biogeochemical Modeling System (GEMS). The GEMS was originally developed for scaling up carbon stocks and fluxes from sites to regions, with measures of uncertainty (Liu et al., 2004). But it can be easily adapted for assessing the impacts of land use and land cover change (LUCC) in general or ecological restoration in particular on soil erosion and deposition, biodiversity, and other ecosystem functions and services (Fig. 3.2).

More specifically, the GEMS relies on a site-scale biogeochemical model to simulate the ecosystem service dynamics. The deployment of the site-scale model

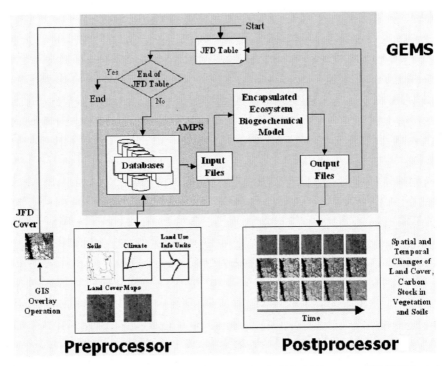

Fig. 3.2 A Diagram of the General Ensemble Biogeochemical Model System, or GEMS (Adopted from Liu et al., 2004)

is based on the spatial and temporal joint frequency distribution (JFD) of major driving variables (e.g., LUCC, climate, soils, disturbances, and management). At the site scale, the GEMS uses stochastic ensemble simulations to incorporate input uncertainty and to quantify uncertainty transfer from input to output. Its description is documented in Liu et al. (2004), and it has already been modified in Zhao et al. (2008) to suit the Chinese context.

The GEMS has four components: a preprocessor, an ecosystem biogeochemical model, an automated model parameterization system (AMPS), and a postprocessor. The preprocessor produces geospatial databases, such as land cover maps and spatial and temporal JFD of major driving variables. The ecosystem biogeochemical model simulates the influences of various factors on ecosystem services dynamically, say, rainfall-induced soil erosion as well as its effect on organic carbon change in soil profiles (Liu, Bliss, Sundquist, & Huntington, 2003). While it was originally built on the Century ecological model (Liu et al., 2003), it is capable of accommodating other modules specializing in certain tasks. The AMPS searches for and retrieves relevant information from various databases according to the keys provided in JFD table, downscales the aggregated information at the map unit level to the field scale using a Monte Carlo approach, and injects the retrieved or assimilated data into the encapsulated ecosystem biogeochemical model by updating its input files. Finally, the postprocessor includes the analysis of simulated results and visualization of the output in forms of spatial maps and graphs.

Below, we elaborate the basic directions and steps in assessing the ecological impacts of soil erosion and water balance, carbon stock and flux, and biological diversity protection.

3.4.1 Soil Erosion and Water Balance

Erosion assessment in China has been mostly conducted either within a small watershed or for the whole nation (Yang & Liang, 2004; Guo, 2003), despite the fact that a regional-level probe is more meaningful (McVicar et al., 2007). But one thing of emphasis is that at the regional level erosion and deposition must be considered simultaneously; otherwise, an inevitable question is: Where has the eroded soil gone? Also, it is at the regional level that the water runoff, scarcity, and other issues can be sensibly elucidated (Li, Qi, Feng, Zhang, & Zhang, 2008). Indeed, even in the broader international context, the link between deforestation vs. flooding/erosion remains elusive. A review conducted by the Center for International Forestry Research (CIFOR, 2005) asserted that evidence did not support the conventional wisdom linking deforestation to large-scale flooding, whereas a new study found that flood risk is correlated with loss of natural forest cover (Bradshaw, Sodhi, Peh, & Brook, 2007). An interesting study notwithstanding, the latter was based on national-level data from 56 developing countries from 1990 to 2000. More work is called for at the sub-national scale with longer data coverage to illuminate whether the link between forests and grassland and floods and erosion is strong enough to justify the large outlay to retain or restore forest and grass covers.

3.4.2 Carbon Storage and Flux

Fang et al., (2001) estimated carbon storage in China's forests over time and found a decline from the beginning of the 1950s to the end of the 1970s due to exploitation, and an increase thereafter during the period of 1980–1998 because of protection and afforestation. Their study considered carbon stored in the forest biomass, but it did not include forest soil carbon or carbon stored in other types of terrestrial ecosystems such as cropland and grassland. Another challenge is the incorporation of the effect of climate change into the assessment. Obviously, there is a long way to go before the spatial and temporal carbon dynamics of all of the ecosystems is known in conjunction with the LUCC and ecological restoration. In our view, work in this direction is critical to our understanding of regional patterns of carbon storage and loss (Zhao et al., 2008); and it will pave the way for land-based carbon accounting, emissions reduction, credits trading, and management practices in a post-Kyoto-Protocol era (Skole, 2007).

3.4.3 Biological Diversity

Our study sites contain thousands of endemic plants and vertebrates, many of which are rare or endangered (Boufford & van Dijk, 1999; Miller, 2004). We can adopt a two-pronged approach to quantifying the effects of past and future LUCC on the regional biodiversity. The first is to use field survey data to relate plant species richness with biophysical variables—forest community type, management status (natural, secondary, and planted), elevation, and aspect. These relationships can then be calibrated in the GEMS to estimate landscape-level plant diversity from geospatial data describing the changes in forest communities and management status over time (Ortega, Elena-Rosello, & Del Barrio, 2004). The second is to assess the effects of LUCC on focal vertebrate species of conservation concern, including the endangered giant panda and black snub-nosed monkey in the southwest and the yak and Mongolian gazette in the northwest. Derived as a function of forest type, management status, slope, and elevation, habitat suitability indices for these species will then be coupled with the historical and future LUCC to map out region-wide changes in habitat for these species and predict the impacts on their populations.

3.5 Assessing Socioeconomic Impacts

In comparison, once data are available, it is much easier to identify a socioeconomic change than to determine how much of it is coming from the treatment effect of a program. So, we will concentrate on how to estimate the treatment effects in this section. Our empirical models should be constructed according to the specific hypotheses to be tested at the individual household, community, or other aggregate levels.

The success of a program depends on its ability to protect the environment, to provide adequate levels of income to participants, and to do so in a cost-effective and sustainable way (Uchida et al., 2005; Zhang et al., 2005). If the program does not target properly, if it does not compensate the participants fairly, and if it cannot keep the participants from reversing their course of action after the end of the program, the fear is that the program may be repeating some of China's early forestry failures (Yin, 1998; Smil, 1993). As such, it is important to evaluate the socioeconomic impacts as reflected in livelihood change, targeting efficiency, and cost effectiveness.

3.5.1 Livelihood Change

Our field knowledge indicates that income and employment are the two most important indicators of livelihood changes caused by the restoration programs. The relative changes of income and employment in various economic sectors are also interesting for detecting the emergence of new opportunities and thus the program sustainability in the long run. In the case of the SLCP, because most of the rural households in the target areas are poor and rely on farming, much of which is on steep slopes or in desertified areas, in order to make the program attractive, it must be able to provide incentives for farmers to participate and to earn enough.

To test whether participating households are at least as well off after the program in the short run, let Y_{it} be the income/employment of household i in sector j at time t, and P the dummy variable representing a households participation choice (1 if participated, 0 if not). Also, let T be another dummy variable indicating the time of observation prior to or after the program introduction (1 if after, 0 if prior to), and X_{it} a set of control variables (e.g., per capita cropland, number of laborers and their education, family size, age and leadership status of the household head) influencing income and/or employment. Then,

$$Y_{ijt} = \alpha_0 + \alpha_1 P_i + \alpha_2 T_t + \alpha_3 (P_i \cdot T_t) + \alpha_4 X_{it} + \varepsilon_{ijt} \tag{3.2}$$

where $\alpha_0, \alpha_1, \alpha_2, \alpha_3$, and α_4 are coefficients to be estimated, and ε is the error term. Again, α_3 will capture the treatment effect.

As to the long-term sustainability of the SLCP, additional analyses can be performed in a similar way. First, we can examine whether farmers are shifting their resources away from cultivation to other productive uses, so that they are increasing the opportunity cost of re-conversion. This may be tested by detecting the changes in livestock activities, off-farm labor, and non-agricultural activities that have occurred since the start of the program. We can also look into the types of trees that are being planted under the program. If households have been able to plant species that allow them to harvest non-timber products, this means that there will be a greater likelihood that the program investment will be long term and generate cash flows in the future.

Similar models can be constructed to determine whether the NFPP has led to livelihood improvement for state employees but worsening for rural households in areas near the natural forests, and how implementing the NFPP has restricted community log production (Chapter 2). Once again, we can relate the income, employment, and other relevant changes with the participation choice, time of observation, and characteristics of the individual agents.

3.5.2 Targeting Efficiency

For the SLCP, one major issue is whether or not the plots with the most suitable features for retirement have indeed been retired. If so, targeting is successful. Since the plots of the participants vary a lot in terms of their productivity and susceptibility to soil erosion, a successful program should also be able to induce households to retire land that has relatively little effect on family income. Given that the main goal of the program is to prevent soil erosion, its designers have made the steepness of the slope one of the main criteria on which plots are selected for inclusion into the program—in southwest China the program targets land with 25° of slope or more to participate. In the northwest the program targets land with 15° of slope or more.

To examine targeting efficiency of the SLCP—the likelihood of certain croplands to be retired—some analysts have estimated a version of the following logit model, in which the probability of land retirement is a function of plot features, household characteristics, and other variables (Xu, Tao & Xu (2004); Uchida, Xu, & Rozelle, 2005). That is, the probability of retirement p_{ij}, a discrete variable indicating whether household i retires plot j (= 1 if retired, 0 if not), is a function of the plot features $Plot_{ij}$ (e.g., slope [<25°, 15~25°, <15°], soil quality [high, average, or low], distance from home [close, or far away]), household characteristics (e.g., size, working members, per capita landholding, age and education of household head), and other variables (e.g., county/township dummies):

$$p_{ij} = \Pr\langle retirement|\, Plot_{ij}, HS_i, D_i\rangle + \varepsilon_{ij} = \Phi\left(\alpha + Plot_{ij}\,\beta + HS_i\,\gamma + D_i\,\theta\right) + \mu_{ij}$$
$$(3.3)$$

where $\Phi(\cdot)$ denotes the cumulative logistic distribution function, α, β, γ, and θ are coefficients to be estimated, and μ is the error term.

The thrust of this model is to determine whether or not it can be statistically shown that plots with high slopes and low yields were indeed targeted in the program, controlling for other household and plot characteristics. A positive and significant coefficient is expected on the slope variable, which suggests that slopes with greater steepness would be more likely selected for participation. In contrast, a negative coefficient on the yield variable would indicate that plots with higher yields would be less likely to be selected for participation.

Because we doubt that the steepness of slope per se is an adequate proxy for erosion severity, however, the above model should be run alternatively. First, we will include the steepness proxy; then, we will replace it with the assessed soil losses of different plots at the beginning of the program. In so doing, we will be able to further infer whether or not the targeting efficiency has been really as high as what was reported and, if necessary, how to enhance that efficiency.

Moreover, the detected forest or grass types, survival, stocking, or growth rates under both the NFPP and SLCP can be regressed against such factors as tenure status, quality and availability of seeds or seedlings, provision of technical service, amount and timing of subsidy delivery, and characteristics of agents. In addition to elucidating the targeting efficiency, findings from this kind of analysis will be useful for demonstrating the implementation efficacy.

3.5.3 Cost Effectiveness

High targeting efficiency may nevertheless not result in cost effectiveness. A policy is cost effective only if it achieves the policy objective at the lowest possible cost. For the SLCP, we can compare the program subsidy with actual grain production by farmers to determine whether and how the land retirement is over- or under-subsidized, and whether other localized practices of implementation can save government funding. The main outlays for the government associated with implementing the SLCP are the set-aside payments that it must make to farmers.

In principle, if the cost effectiveness of China's conservation set-aside program is to be optimized, both environmental and productive heterogeneity need to be considered. That is, the cost effectiveness would be greatest if payments were indexed on the basis of each plot's slope (choosing the steepest ones, which might generate the greatest environmental benefit) and yield history (choosing the lowest yielding one, which would require the lowest payment to cover its opportunity cost). Under the current mechanism, however, the government offers only two levels of compensation for participation in the two large basins of the Yangtze River and the Yellow River, with a difference in the grain subsidy only (Chapter 1). The program has not closely matched payments to maximize the environmental benefit and minimize payments. Thus, better targeting could have reduced costs to the government as well as to the farmers.

To see how well the program has produced environmental benefits in a cost-effective way and to determine which factors cause the variation of cost effectiveness, we postulate that the cost differential ΔC (compared to the baseline) for agent i in year t is determined by the specific practice M, institutional arrangement A (coordination and monitoring), community involvement I (whether participated and what decided), characteristics of the agents X, and locational dummy D:

$$\Delta C_{it} = \beta_0 + \beta_1 A_i + \beta_2 M_t + \beta_3 I_t + \beta_4 X_{it} + \beta_5 D_i + \nu_{it} \qquad (3.4)$$

where β_0, β_1, β_2, β_3, β_4, and β_5 are coefficients to be estimated, and ν is the error term. Formulated and estimated properly, the significance of these coefficients will reveal both the degree of cost effectiveness and its determinants.

For the NFPP, we can investigate whether public investments in forest protection, regeneration, and management are justified in relation to the expenditures of local best practices. It has been argued that if some of the activities could be contracted out to non-state workers, instead of having state employees perform the activities, a large portion of the outlays would be saved (Yin, Xu, Li, & Liu, 2005). By comparing the current payment rates with what nearby farmers would offer and running a similar switch regression against the institutional, agent, and other features, we can shed a great deal of light on the above question.

3.6 Data Generation and Acquisition

The analytic tasks outlined above require a large amount of comprehensive and accurate data, and data availability and reliability are important to a successful assessment. However, data are generally not easily available, accessible, or of high quality in China. Here, we describe our strategies to overcome these difficulties.

First, by adopting a stratified sampling technique, we can gather the needed social economic data from a survey. That is, we pick some representative counties (of erosion and desertification control) and state forest bureaus (of natural forest protection) from the relevant regions first; then, we choose certain townships, forest farms, households, and state employees to administer the necessary questionnaires, including the status and contract terms of program participation, socioeconomic activities, perceptions, and characteristics of individuals and local organizations. These survey efforts should result in a large number of observations for executing the socioeconomic research tasks.

Second, we will derive the attributes and dynamics of the LUCC information for the selected study sites from interpreting satellite images to discern the changes resulted from implementing the restoration programs. Given that China's official LUCC statistics are notoriously inaccurate or unavailable and that China has few field stations for biological and hydrological observations (Zhao et al., 2008), this strategy is especially beneficial. And advances in geospatial science and GIS application have made it much easier to measure and analyze LUCC and ecosystem conditions (Lambin, Geist, & Lepers, 2003; Foley et al., 2005). The primary source of the spatial information is the Enhanced Thematic Mapper (ETM) images for the period of 1997–2005. Consistent with the common land-use classification system, regional land uses will be classified into six primary categories (cropland, woodland, grassland, urban and other built-up areas, water, and other) and 18 secondary categories (Liu et al., 2003).

Third, the bulk of the other biophysical data can be collected from existing sources. For instance, soil data can be obtained from a national database with multilayer information at a 10×10 km resolution, including soil texture, bulk density, organic matter content, and wilting point and field capacity for each layer (Shi & Yu, 2002); watershed drainage classes can be found from a GIS-based moisture index by integrating topography, flow accumulation, curvature, and water holding capacity with digital elevation model and soil texture data (Iverson, Dale, Scott, & Prasad, 1997); climate data, containing monthly minimum, mean and maximum temperature, and mean precipitation, can be obtained from a country-wide database of $0.1° \times 0.1°$ resolution (Fang et al., 2001); and nitrogen deposition from wet and dry sources at the county level can be found from the literature (Li & Frolking,

2006). If necessary, these and other data can be updated and expanded. To validate data and calibrate simulations, ground truthing and field experiments (e.g., land-use practices vs. soil erosion or carbon cycling) can be undertaken.

Finally, we will strive to build longitudinal data of socioeconomic and ecological nature with longer series. As noted, the assessments performed so far have data for a few years, which may have captured only the early, partial effects of a program. Further, it is desirable if we can go back to the 1980s or even to an earlier period of time to trace out the land use change trajectory and identify the patterns and turning points, so that we will be able to put the LUCC in a historic perspective and relate it with the potential drivers more clearly. Additionally, our data must be generated in line with assessment at multiple scales. That is, micro-level data will allow us to examine how local farmers respond to the programs and what the induced consequences are, but this type of data can hardly capture LUCC on a larger scale and is incompatible with evaluating the accompanying environmental impacts. Moreover, certain ecological consequences and their spatial heterogeneity can be examined only with data covering a large geographic scale.

3.7 Synthesis and Actions

In this chapter, we began with a claim that the assessment of China's ecological restoration programs has so far not been well integrated, with more attention devoted to the short-term socioeconomic impacts. Without an integrated and more balanced assessment, however, we will not be able to provide effective and timely answers to such key questions as: Have these programs been effectively implemented, and if so, how? Have they affected the socioeconomic situations and the ecological processes and if so, how? Therefore, we set out to articulate and outline our methodology for an integrative assessment of these programs, based on our own research experience.

First, we argued that an integrative assessment of the programs should: (1) embrace both socioeconomic and environmental impacts as well as the effectiveness of their implementation; (2) be conducted at multiple scales with a regional focus and spatially explicit longitudinal data; (3) engage active interdisciplinary collaboration; and (4) connect science with policy. Then, we described in detail our general approach and specific models to quantify the environmental and socioeconomic impacts induced by implementing the restoration programs, and addressed how to overcome the challenges in generating all of the needed data for executing these empirical tasks.

We believe that the methodology that we have put forth is novel in integration and feasible in execution, and that the research agenda that we have outlined is essential and practical. It is worth reiterating that, in order to implement the methodology and fulfill the agenda, the following specific actions must be taken: (1) build capacity of a multidisciplinary team with expertise in landscape ecology, social science, remote sensing and GIS, and climate change; (2) develop comprehensive datasets of spatial, census, and survey sources and robust treatment effect and ecological models; (3)

integrate analytic efforts at the individual, community, and regional scales; and (4) collaborate with Chinese scholars and practitioners.

The later chapters of this book are empirical studies that feature, one way or another, the methodology and agenda that we have proposed. We hope that after having browsed those chapters, the readers will come to an agreement with us in that these empirical efforts in combination represent a major advancement of our knowledge of the implementation effectiveness and impact significance of China's ecological restoration programs. Moreover, we hope that the readers will concur with us in that a broader adoption of this methodology and a further execution of this agenda will lead to a much improved scientific understanding of the program impacts and a greatly enhanced institutional capability of implementing similar programs in and outside of China.

Looking ahead, this methodology will take us a long way into measuring and evaluating ecosystem services. In their "call to ecologists," Kremen and Ostfeld (2005) stated that, in order to manage ecosystem services in the future, we need to better understand their underlying ecology, and use it to develop better market signals for ecosystem services, create better strategies and policies for their conservation and sustainable use, evaluate the tradeoffs between policies and practices that promote different services, and design management and conservation plans for services at the whole-system level. Foley et al. (2005) also noted that society faces the challenge of developing strategies that reduce the negative environmental impacts of land use while maintaining socioeconomic benefits and sustainability policies that must assess and enhance the resilience of different land-use practices. We are excited to put these good concepts into effect by assessing the impacts of China's ecological restoration programs with an adequate framework.

Acknowledgements This work was supported by the U.S. NSF project (0624018). The authors are grateful for Erin Shi and Victoria Hoelzer-Maddox for their assistance.

References

Angrist, J., & Krueger, A. (1999). Empirical strategies in labor economics (chap. 23). In O. Ashenfelter & D. Card (Eds.), *The handbook of labor economics* (Vol. III). Amsterdam: North Holland.

Bennett, J., Wang, X., & Zhang, L. (Eds.). (2008). *Environmental management in China: Land use management*. Cheltenham: Edward Elgar.

Boufford, D. E., & van Dijk, P. P. (1999). South-central China. In R. A. Mittermeier, N. Meyers, P. Robles Gil, & C. G. Mittermeier (Eds.), *Hotspots: Earth's biologically richest and most endangered terrestrial ecoregions* (pp. 338–351). Mexico: CEMEX.

Boyd, J. (2007). The endpoint problem. *Resources* (Spring) 26–28.

Bradshaw, C. J. A., Sodhi, N. S., Peh, K. S., & Brook, B. W. (2007). Global evidence that deforestation amplifies flood risk and severity in the developing world. *Global Change Biology, 13*, 2379–2395.

CIFOR. (2005). *Forests and floods: Drowning in fiction or thriving on facts?* Bangkok, Thailand.

Clark, W. C. (2007). *Sustainability science: An emerging interdisciplinary frontier. The Rachel Carson distinguished lecture series*. East Lansing, MI: Michigan State University.

Conservation International. (2002). *Biodiversity hotspots*. http:// www.biodiversityhotspots.org/ xp/hotspots/China.

Fang, J. Y., Chen, A. P., Peng, C. H., Zhao, S. Q., & Ci, L. J. (2001). Changes in forest biomass carbon storage in China between 1949 and 1998. *Science, 292*, 2320–2322.

FAO. (2006). *Global Forest Resources Assessment 2005.*Rome: The Food and Agricultural Organization.

Foley, J. A., DeFries, R., Asner, G. P., Barford, C., Bonan, G., Carpenter, S. R., et al. (2005). Global consequences of land use. *Science, 309*, 570–574.

Forest and Grassland taskforce (1/ 2003). *In Pursuit of a Sustainable Green West* (Newsletter).

Guo, Y. S. (2003, October 27–29). *Establishing information system of soil conversion in China.* National water conservation information technology conference. Shanghai.

Heckman, J., Lalonde, R., & Smith, J. (1999). The Economics and Econometrics of Active Labor Market Programs. In O. Ashenfelter & D. Card (Eds.), *The Handbook of Labor Economics* (Vol. 3A). Amsterdam: North Holland.

Imbens, G., & Woodridge, J. (2007). *Mini-Course of What's New in Econometrics.* Cambridge, MA: The National Bureau of Economic Research Summer Institute.

IPCC (Intergovernmental Panel on Climate Change). (2007). *Summary for policymakers. Fourth Assessment Synthesis Report* (http://www.ipcc.ch).

Iverson, L. R., Dale M. E., Scott C. T., & Prasad A. (1997). A GIS-derived integrated moisture index to predict forest composition and productivity of Ohio forests (USA). *Landscape Ecology, 12*, 331–348.

Kremen, C., & Ostfeld, R. S. (2005). A call to ecologists: measuring, analyzing, and managing ecosystem services. *Frontiers in Ecology and Environment, 3*(10), 540–548.

Lambin, E. F., Geist, J. H., & Lepers, E. (2003). Dynamics of land-use and land-cover change in tropical regions. *Annual Review of Environment and Resources, 28*, 205–241.

Lee, M. J. (2005). *Micro-econometrics for policy, program, and treatment effects.* Oxford, UK: Oxford University Press.

Li, C. B., Qi, J. G., Feng, Z. D., Zhang, X. W., & Zhang, F. (2009). Process-based soil erosion simulation on a regional scale: the effect of ecological restoration in Chinese Loess Plateau. *Environmental Management* (submitted).

Li, C. S., & Frolking, S. (2006). *Water drainage classification.*Complex Systems Research Center, Institute for the Study of Earth, Oceans, and Space, Morse Hall, University of New Hampshire, Durham, New Hampshire, USA. http://eos-webster.sr.unh.edu/data_guides/china_dg.jsp

Liu, S. G., Bliss, N., Sundquist, E., & Huntington, T. G. (2003). Modeling carbon dynamics in vegetation and soil under the impact of soil erosion and deposition. *Global Biogeochemical Cycles, 17*(2), 1074, doi: 10.1029/2002GB002010.

Liu, S. G., Loveland T. R., & Kurtz, R. M. (2004). Contemporary carbon dynamics in terrestrial ecosystems in the Southeastern plains of the United States. *Environmental Management, 33*, S442–S456.

McVicar, T. R., Li, L. Van Niel, T. G., Zhang, L., Li, R., Yang, Q., et al. (2007). Developing a decision support tool for China's re-vegetation program: Simulating regional impacts of afforestation on average annual streamflow in the Loess Plateau. *Forest Ecology and Management, 251*(1–2), 65–81.

Millennium Ecosystem Assessment (MA).(2005). *MA Findings* (www.millenniumassessment. org).

Miller, D. (2004). Searching for grass and water: grazing land ecosystem sustainability and herder;s livelihood in western China. Unpublished report 1–63.

Ortega, M., Elena-Rosello, R., & Del Barrio, J. M. G. (2004). Estimation of plant diversity at the landscape level: A methodological approach applied to three Spanish rural areas. *Environmental Monitoring and Assessment, 95*, 97–116.

Shi, X. Z., & Yu D. S. (2002). A framework for the 1:1,000,000 soil database of China. In proceedings of the 17th World Congress of Soil Science. Bangkok, Thailand.

Skole, D. (2007). *Carbon2Markets: Next generation technologies in carbon market development.* (working paper, Michigan State University).

Smil, V. (1993). Afforestation in China. In A. Mather. (Eds.) *Afforestation: Policies, Planning and Progress.* London, UK: Belhaven Press.

Turner II, B. L., Lambin, E. F., & Reenberg, A. (2007). The emergence of land change science for global environmental change and sustainability. *PNAS, 104*(52), 20666–20671.

Uchida, E., Xu, J. T., & Rozelle. S. (2005). Grain for Green: cost-effectiveness and sustainability of China's conservation set-aside program. *Land Economics, 81*(2), 247–264.

Uchida, E., Xu, J. T., Xu, Z. G., & Rozelle, S. (2007). Are the poor benefiting from China's land conservation program? *Environment and Development Economics, 12*, 593–620.

Wooldridge, J. M. (2000). *Introductory econometrics: A modern approach.* Mason, OH: South-Western College Publishing.

Xu J. T., Tao, R., & Xu, Z. G. (2004). Sloping land conversion: Cost-effectiveness, structural adjustment, and economic sustainability. *Quarterly Economics, 4*(1), 139–162 (in Chinese).

Xu, J. T., Yin, R. S., Li, Z., & Liu, C. (2006). China's ecological rehabilitation: Progress and challenges. *Ecological Economics, 57*(4), 595–607.

Yang, Z. S., & Liang, L. H. (2004). Soil erosion under different land use types and zones of the Jinsha River basin in Yunnan province. *Journal of Mountain Science, 1*, 46–56.

Yin, R. S. (1998). Forestry and the environment in China: the current situation and strategic choices. *World Development, 26*(12), 2153–2167.

Yin, R. S., Xu, J. T., Li, Z., & Liu, C. (2005). China's ecological rehabilitation: Unprecedented efforts, dramatic impacts, and requisite policies. *China Environment Series, 6*, 17–32.

Zhang, S. Q., Swanson, T., & Kontoleon, A. (2005). *Impacts of compensation policies in reforestation programs.* (Report of the Environment and Poverty Programme to China Council for International Cooperation in Environment and Development.)

Zhao, S. Q., Yin, R. S., Liu S. G., Li, Z. P., Deng, Y. L., Tan, K., et al. (2009). Quantifying terrestrial ecosystem carbon dynamics in the Jinsha watershed of Upper Yangtze River from 1975 to 2000. *Environmental Management* (submitted).

Chapter 4
Land Cover Changes in Northeast China from the late 1970s to 2004

Zhangquan Shen, Runsheng Yin, and Jiaguo Qi

Abstract This chapter presents a quantitative analysis of land cover changes in northeast China from the late 1970s to 2004 using remote sensing and the Geographic Information System. Land covers are mapped into six classes and nine sub-classes from multi-temporal Landsat MSS, TM and ETM+ images and SRTM DEM data. It is found that while forestland and wetland were greatly reduced until 2000 due to farming expansion and urbanization, spurred by population growth, their decline trends have been revered most recently. Meanwhile, built-up land has kept increasing. Further, the land cover changes occurred primarily in areas with low elevation and gentle slope. These results suggest that the forest and wetland protection and restoration projects have taken effect. However, there remains a long way to go before the ecosystems are greatly recovered and can function in the way that society expects.

Keywords Ecological restoration · Land cover change · Landsat MSS/TM/ETM+ · Digital elevation model · Northeast China

4.1 Introduction

In recent decades, land use and land cover change (LUCC) has become an important part of global change research because of its interaction with climate, ecosystems, biogeochemical cycles, and human activities (Xiao et al., 2006). The work of many investigators has shown that LUCC affects the global system in various ways, including atmospheric composition, regional climate, soil quality, hydrology, and biodiversity (Xu, Liu, An, Chen, & Yan, 2007; Turner, Lambin, & Reenberg, 2007). Humans are the main force behind the global conversion of land cover (Kuemmerle,

Z. Shen (✉)
College of Environmental and Resource Sciences, Zhejiang University, Hangzhou 310029, P. R. China; Center for Global Change and Earth Observations, Michigan State University, East Lansing, MI48823, USA
e-mail: zhqshen@zju.edu.cn

Radeloff, Perzanowski, & Hostert, 2006), and their ecological and environmental influence has been steadily increasing.

Remote sensing can provide spatially consistent data sets that cover large areas with both high spatial detail and temporal frequency, and it is accepted as a "unique view" of the spatial and temporal dynamics of the LUCC processes. Therefore, remote sensing techniques have been widely used in detecting and monitoring land cover change at various scales (Xiao et al., 2006; Keuchel, Naumann, Heiler, & Siegmund, 2003). The goal of this chapter is to assess the land cover change in northeast China using remote sensing and GIS techniques. While we will focus on the change since the late 1990s when various forest and wetland protection and restoration projects were launched, we intend to trace the regional land cover change back to the late 1970s to put it in the proper context.

Due to its broad and repetitive coverage, Landsat has provided the scientific community with valuable imagery that can be utilized for detecting terrestrial land cover conditions and tracking vegetation dynamics, agricultural activity, urban growth, and surface hydrology (Alrababah & Alhamad, 2006). Further, its thermal band measures the emission of energy from the earth's surface, and it can thus be used as an indicator of land cover type based on recorded temperatures (Southworth, 2004). The post-classification method of change detection, which compares two or more separately classified images of different dates, is commonly used (Dewidar, 2004; Yuan, Sawaya, Loeffelholz, & Bauer, 2005).

China's tremendous population explosion, economic growth, and urbanization have caused major problems of resource depletion and ecosystem degradation (Liu & Diamond, 2005; Yin et al., 2005). To tackle these problems, the government has recently initiated a series of ecological restoration programs (Xu et al., 2006). It is interesting and important to detect the induced land cover change of these programs and to relate the change to the policy initiatives in a timely manner. In so doing, not only will it make it possible for the science community to assess the potential impacts of these initiatives on ecosystem functions, but it will also enable us to provide much needed feedback to the agencies regarding the effectiveness of their policies.

While there have been many studies of the LUCC in China, most of them dealt with changes before 2000 (Jiang, 2002; Tang, Wand, & Zhang, 2005; Wang et al., 2005; Xiao et al., 2006; Xu et al., 2007; Zha, Liu, & Deng, 2008). The LUCC in China after 2000 is rarely examined. In order to examine this period, it is necessary for us to acquire and interpret images available for the most recent past. In this chapter, we select ten counties in northeast China as a study area because of its significance as a primary natural forest and wetland region. Our results show that while forestland and wetland were greatly reduced until 2000 due to farming expansion and urbanization, driven by population growth, their decline trends have been reversed most recently. Moreover, the land cover changes occurred mainly in areas with low elevation and gentle slope. These findings suggest that the forest and wetland protection and restoration projects have been effective. However, there is a long way to go before the ecosystems are greatly recovered and can function in a manner that society expects.

4.2 Study Area

The study area is located in Heilongjiang province of northeast China. It includes 10 counties with a total area of about 29,000 km^2, and ranges from 45° 32′ 18″ N to 47° 45′ 8″ N latitude and from 128° 14′ 24″ E to 132° 33′ 31″ E longitude. Its elevation is between 23 and 1,307 m above sea level. The northeast part belongs to the Sanjian Plain and is the lowest part in the study area, whereas the middle and southern parts are hilly and mountainous (see Fig. 4.1). This area has a temperate, humid to sub-humid continental monsoon climate. The annual temperatures are 21–22°C (average maximal), 1.4–4.3°C (mean), and −18°C (average minimal). Mean annual precipitation is 500–600 mm, and 80% of all rainfall occurs between May and September. The frost-free period is 120–140 days.

Prior to 1950, the region was endowed with primary natural forests and wetland. From the 1950s to the late 1990s, however, a lot of the forestland and wetland was cleared for farming and timber (see Chapter 1). Since then, the central government has invested heavily in protecting the natural resources and restoring the ecosystems. While it is true that the environmental conditions have improved significantly in recent years, little has been done to assess to what extent the resource conservation projects have made a difference.

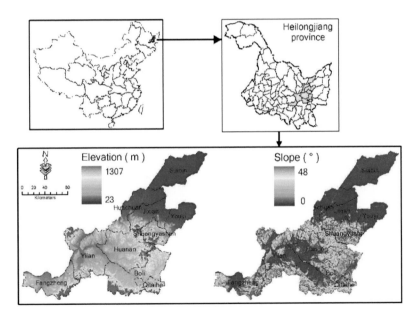

Fig. 4.1 Location of the Study Area and Its Elevation and Slope

4.3 Materials and Methods

4.3.1 Landsat Imagery

The Landsat images used in this study cover four points of time from the late 1970s to 2004. They include one set of MSS images for the late 1970s (hereafter, 1977), two sets of TM images for around 1990 (hereafter, 1990) and 2004, and one set of ETM+ images for around 2000 (hereafter, 2000). Each set has three images to cover the entire study area. The path, row, and acquired date for each scene are listed in Table 4.1. Notice that due to quality concerns, images for a given year may not be available; a common practice in this circumstance is to assemble them around a given year as closely as possible (Xiao et al., 2006; Jarvis, Reuter, Nelson, & Guevara, 2006). Also, notice that because of the quality deterioration of ETM+ images for 2004, TM images are used.

The MSS, TM, and ETM+ images for the first three points of time were downloaded from the Global Land Cover Facility site (http://glcf.umiacs.umd.edu), and they were georeferenced and rectified by the GLCF to UTM projection zone 52 and WGS84 datum with a spatial resolution of 57, 28.5, and 28.5 m, respectively. Their TM6 images were re-sampled to match the resolution of other bands using the nearest-neighbor method. The TM images for 2004 were ordered from China Remote Sensing Satellite Ground Station, and they were geo-encoded and matched to the ETM+ images one by one with a total RMS error of less than 0.5 pixels using the image-to-image registration method. Then, they were re-sampled to 30 m using the nearest-neighbor method. Nearly all of the Landsat images are free of cloud. Finally, all sets of images were masked using the boundary of study area.

Table 4.1 Brief Information for Landsat Imagery Used in the Study

Date	Late 1970s	Around 1990	Around 2000	Around 2004
Type of sensor	MSS	TM	ETM+	TM
Path and row	124–27	115–27	115–27	115–27
	124–28	115–28	115–28	115–28
	125–28	116–28	116–28	116–28
Acquired date	24/08/1978	25/06/1991	12/08/2000	12/06/2004
	27/05/1976	02/09/1993	12/08/2000	12/06/2004
	03/07/1976	23/05/1994	03/06/2001	22/08/2004

4.3.2 Topographic Data

Topography is a fundamental geophysical variable that contains valuable information about the geodynamic and climatic history of a region (Coblentz & Riitters, 2004). It can be used in ecology, hydrology, agriculture, and many other fields as a means of explaining and predicting processes through modeling. It has also been used as a supplementary source of information for land cover classification.

In addition, slope and aspect data produced from a coarse spatial resolution digital elevation model (DEM) can be integrated with multi-spectral data for land cover analysis (Nguyen, Atkinson, & Lewis, 2005).

Detailed and accurate DEM information can be generated from remote sensing systems, such as SPOT and ASTER, and finer differences in elevation can be obtained with airborne laser-scanners. NASA recorded most of the world's topography in the Shuttle Radar Topographic Mission (SRTM), and these data have been available for nearly the entire globe. An evaluation test indicated that if only cartography with scales above 1:25,000 (i.e., 1:50,000 and 1:100,000) is available, it is better to use the SRTM DEMs. The SRTM DEMs can be used for terrain derivatives (slope, aspect, landscape classifications, etc.) as well (Jarvis, Rubiano, Nelson, Farrow, & Mulligan, 2004). However, some regions may miss data because of a lack of contrast in the radar image due to presence of water, or excessive atmospheric interference. These data holes are especially concentrated around rivers, lakes, and steep areas. This non-random distribution of holes impedes the potential use of SRTM data and has been the subject of a number of innovative algorithms for "filling-in" through spatial analysis techniques. In this study, the post-processing 3 arc second SRTM DEM data were acquired from CGIAR-CSI (the Consortium for Spatial Information of the Consultative Group for International Agricultural Research) web site by a geographic projection with the WGS84 horizontal datum and EGM96 vertical datum (Jarvis et al., 2006). The non-data holes in the original DEM were filled with the help of available auxiliary DEM data through an interpolative technique within an Arc/Info AML model, from which the slope information was derived. The DEM and slope data were masked using the region boundary of study area.

4.3.3 Pre-processing and Classification

A land-use map at a scale of 1:500,000 was obtained for the selection of training data and validation of classification results. The map covers the whole Heilongjiang province around 1995. Even though it was developed following the old version of the Chinese land-use classification system (Chinese National Land Resource Survey Committee, 1984) and its resolution is a bit coarser, it is still very helpful.

The whole image pre-processing included two steps. First, MSS 4-7 or TM/ETM 1-5,7 were transformed by Principal Component Analysis (PCA) method and the first n principal components that accounted for over 98% of the variance in the images based on Eigen-value analysis were extracted. Second, the extracted PC images, elevation, slope and thermal band (only for TM and ETM images) were stacked together. Before classifying land cover, a two-tier hierarchical classification scheme was set up with the assistance of long-term field knowledge gained from geography, vegetation, and land cover in northeastern China. There are six primary classes: farmland, forestland, grassland, built-up land, water body, and unused land. Farmland, forestland, and grassland are further divided into certain sub-classes (Table 4.2).

Table 4.2 Description of the Land Cover Classification System Used in this Study

Level 1 class	Level 2 class	Description
Farmland	Paddy land	Farmland mainly used for growing paddy rice and lotus roots with guaranteed water source or irrigation facilities, including paddy field rotated with dryland crops.
	Dry land	Rain-fed farmland without irrigation; dry farmland with irrigation; land mainly for growing vegetables; and fallow land.
Forestland	Dense	Natural or plantation forest with canopy cover >40%.
	Sparse	Natural or plantation forest with canopy cover ≤40%.
Grassland	Dense	Natural or artificial grassland with canopy cover >20%.
	Sparse	Natural or artificial grassland with canopy cover ≤20%
Bodies of Water		Natural water system and land for irrigation facilities, including river, lake, pond, glacier, reservoir, etc.
Built-up land		Residence, transportation network, and other buildings, including land for urban occupation, etc.
Unused land		Land not yet used, including those difficult to be used such as desert, gobi, salina, barren soil, bare rock, wetland, etc.

The widely used supervised classification method, Maximum Likelihood Classification (MLC), was employed. That is, the ancillary land cover map was overlaid on the images, and the training site data were collected by means of the on-screen selection of the polygons. The NDVI (Normalized Difference Vegetation Index) images derived from the original images were used to help the identification of sub-classes of forestland and grassland based on the MLC. Due to satellite instruments and other considerations, the images in each set were classified individually. The classified results of images in each set were then combined for further analysis.

The resultant land cover maps were assessed for accuracy in ERDAS IMAGINE ver. 9.1. About 1,500 pixels were randomly selected from each set of classified outcomes in the study area. A confusion matrix was first generated, and then the producer's and user's accuracy as well as the Kappa coefficient for each class and the whole image set were derived.

In determining land cover change, a cross-tabulation detection method was adopted. A change matrix was produced, showing the overall land cover changes and gains and losses in each class. And the change matrix reveals the main types of changes in the study area (Wang et al., 2005).

4.4 Results

The derived accuracy measures for each land cover class and the whole image set are summarized in Table 4.3. While some classes have a relatively low degree of classification accuracy, those major ones, such as forestland and farmland, have a very high degree of accuracy. Further, the overall degree of accuracy is quite high.

Table 4.3 Summary of Accuracy Assessment for Image Classification

Level 1 class	Producer's accuracy (%)				User's accuracy (%)				Kappa coefficient			
	Late 1970s	Around 1990	Around 2000	Around 2004	Late 1970s	Around 1990	Around 2000	Around 2004	Late 1970s	Around 1990	Around 2000	Around 2004
Farmland	94.93	87.54	91.30	91.77	90.35	91.07	89.15	93.43	0.8295	0.8521	0.7578	0.8612
Forestland	95.87	96.21	97.57	93.54	96.03	90.49	92.45	94.84	0.9315	0.8355	0.8920	0.9218
Grassland	30.00	70.00	48.57	21.05	41.38	71.79	77.27	47.06	0.3977	0.7021	0.7673	0.4568
Bodies of water	100.00	83.33	93.55	96.88	95.00	88.24	96.67	93.94	0.9494	0.8809	0.9660	0.9381
Built-up	44.44	74.36	68.57	81.67	100.00	67.44	63.16	52.69	1.0000	0.6657	0.6135	0.5072
Unused land	72.85	76.47	32.14	69.57	82.09	88.14	57.45	62.34	0.8008	0.8695	0.5492	0.6052
Overall	**91.13**	**88.87**	**87.87**	**89.27**	**91.13**	**88.87**	**87.87**	**89.27**	**0.8561**	**0.8285**	**0.7934**	**0.8233**

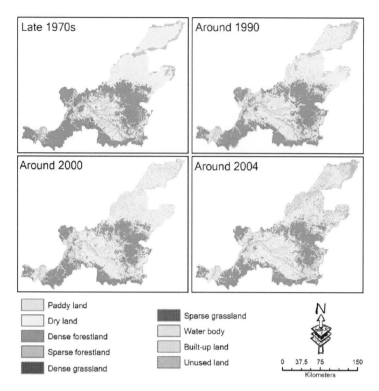

Fig. 4.2 Maps of Land Covers in the Study Area

So, it can be said that the results are reasonably accurate (Alrababah & Alhamad, 2006).

The land cover maps for the four points of time are shown in Fig. 4.2. The overall land cover changes from the late 1970s to 2004 are illustrated in Fig. 4.3. It can be seen that farmland and forestland are the two largest land cover classes in the study area. Farmland took up 46.13–56.75% of the total area, and forestland accounted for 32.69–41.31%. But built-up land had the largest rate of change in the whole study period. The area of farmland increased steadily from the late 1970s to 2000, and then it declined from 2000 to 2004.

From Table 4.4, it is found that the expansion of farmland came mostly from forestland, unused land, and grassland. While the dominant part of the farmland was dry land, the proportion of paddy land increased from the late 1970s to 2000 and then stabilized. The unused land, more precisely wetland in this area, was reclaimed and converted to paddy land. The largest proportion of farmland is located in the area of slopes less than 2° and elevations less than 200 m, and it is noticeable that cropland in this range declined from 2000 to 2004, because of the conversion to built-up land, grassland, or unused land (Figs. 4.4 and 4.5).

The trend of change for forestland from the late 1970 to 2000 was opposite to that of farmland. That is, while farmland increased significantly, forestland declined

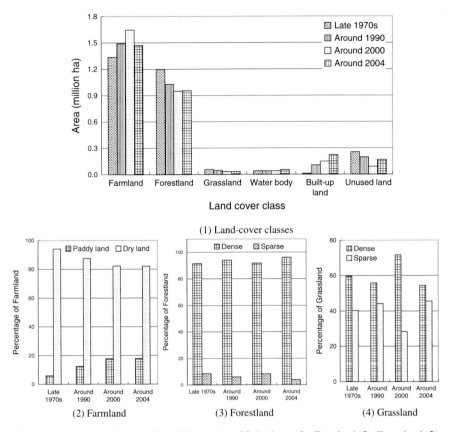

Fig. 4.3 Areas of Land Covers (**1**) and Proportion of Sub-classes for Farmland (**2**), Forestland (**3**), and Grassland (**4**)

rapidly. But forestland slightly increased from 2000 to 2004. Conversion of forestland to farmland was the major driver for the decrease of forestland, as shown in Fig. 4.3. The proportions of sub-classes in forestland were stable from the late 1970s to 2000, and then changed slightly. The dominant sub-class of forestland is dense forest, and its proportion was over 91% in the entire study period. The forest restoration and conservation projects launched in the late 1990s might be a cause for the slight increase of forestland from 2000 to 2004. The forestland was mostly located in the zone of medium slope (2–15°) and elevation (200–500 m), and its change occurred in the zone with low to medium slope (less than 15°) and elevation (less than 500 m); this trend of change coincided with the overall change of forestland in the flat and low elevation area (less than 2° and 200 m), and it was stable in the areas of steep slope (more than 15°) and high elevation (over 500 m).

The percentage of grassland was very low in this area and it had the similar trend of change to forestland. The major conversion for grassland was between farmland, forestland, and itself. Grassland was located mostly in mountainous areas and inside or near forest before 2000, but its major part shifted to low altitude area and plains,

Table 4.4 Land Cover Transition over Time in the Study Area (Unit: ha)

| | Land cover classes | Late 1970s | | | | | |
		Farmland	Forestland	Grassland	Water bodies	Built-up land	Unused land
Around 1990	Farmland	1,158,357	143,955	40,568	6,375	2,201	134,951
	Forestland	32,888	980,366	5,667	463	654	2,679
	Grassland	946	41,699	2,445	109	19	562
	Water bodies	3,721	1,208	299	28,217	10	6,523
	Built-up land	72,150	12,972	4,785	683	7,288	8,138
	Unused land	68,327	15,914	2,988	4,356	97	102,804
		Around 1990					
Around 2000	Farmland	1,327,921	104,707	26,938	4,661	49,210	134,152
	Forestland	31,009	895,773	12,659	65	3,322	6,192
	Grassland	6,927	17,185	4,903	86	950	1,135
	Water bodies	4,482	508	151	31,442	844	2,225
	Built-up land	72,257	7,365	579	3,194	49,794	13,573
	Unused land	46,321	2,090	566	597	2,025	37,370
		Around 2000					
Around 2004	Farmland	1,273,770	92,105	12,318	1,963	53,835	35,995
	Forestland	101,503	832,828	13,917	236	4,819	2,780
	Grassland	24,310	4,571	183	15	2,043	665
	Water bodies	11,946	541	202	33,128	5,543	3,014
	Built-up land	120,873	16,508	4,310	1039	71,973	7,663
	Unused land	114,634	1,705	212	3,145	8,499	38,780

and was converted from farmland by 2004. The cause of the shift might be the policy of protecting wetland and grassland since late 1990s. Most of the grassland was located below 500 m and the main change for grassland occurred in the area of low elevation (less than 200 m) and gentle slope (less than 2°). Likewise, nearly all bodies of water were located in flat areas (less than 2°) below 200 m.

From the late 1970s to 2004, the area of built-up land increased about 20 times due to the population growth and life style change. Most of the expanded built-up land came from farmland, which is easy to distinguish on the land cover maps

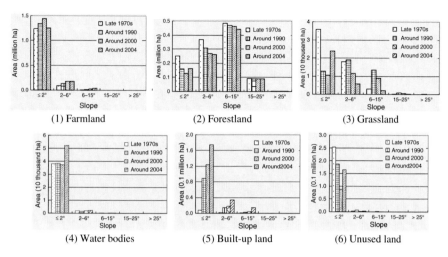

Fig. 4.4 Distribution of Land Covers in Different Slope Zones over Time

(Fig. 4.2). The main part of the built-up land was in the area of low elevation (less than 200 m) and gentle slope (less than 2°). The unused land, or wetland, declined from the late 1970s to 2000, and increased thereafter. The cause for the decline was that the wetland was reclaimed as farmland to meet the increasing demand of food and other products, and the recent increase might be due to the wetland protection policy (SFA, 2007). Most of the unused land was in areas below 200 m and with low slope (less than 2°).

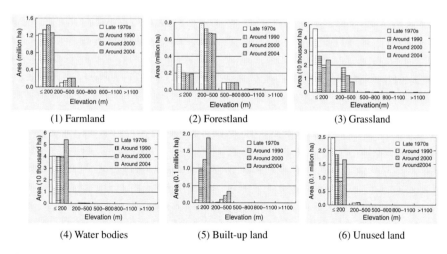

Fig. 4.5 Distribution of Land Covers in Different Altitudinal Zones over Time

4.5 Conclusions and Discussion

Using multi-temporal Landsat imagery and GIS-related techniques, this study has produced fairly accurate land cover maps for ten counties in northeast China, which is one of the focal areas of the country's recent forest and wetland protection and restoration initiatives. These maps enabled us to assess the land cover characteristics and changes over time. Our results indicate that from the late 1970s to 2004, the study area experienced some significant land cover changes, induced by human activities, agricultural policy, and ecological restoration. Year 2000 was a turning point of the land use trends. Before that time, farmland expanded persistently by encroaching on the natural forests, grassland, and wetland, due to population increase and agricultural growth. This led to the depletion of natural resources and degradation of ecosystems as well as reduced economic benefits to farming (Tong, Hall, & Wang, 2003). However, since the forest restoration and wetland protection projects were launched in the late 1990s, the declining trend of forestland was reversed and a large part of the farmland was converted to forestland and wetland. Also, built-up land experienced a continuous increase in the study area, which is typical across the country, given the demographic and economic trends.

While we are satisfied with the accuracy of our results, they can be improved. First, although the terrain and thermal information was considered in the classification and the spectral similarity of land cover classes caused errors, more adequate contextual knowledge should be considered in the classification in order to tackle the "confused" pixels and minimize the error. This will lead to a more accurate assessment of the land cover change. Second, ecological, political and economic factors should be incorporated into the driving forces analysis of LUCC. Finally, taking year 2000 as a turning point, we may derive more detailed land cover information before and after it so as to benefit the analysis of land cover change and its causes and effects. These ideas are being pursued currently.

Acknowledgments This research was supported by the U.S. National Science Foundation as part of the ongoing research (An Integrative Impact Evaluation of China's Ecological Restoration Programs, #0624018) at Michigan State University. The authors are grateful for the comments and suggestions made by participants of the International Symposium on Evaluating China's Ecological Restoration Programs, held on October 19, 2007, in Beijing. They appreciate Brain Walters, Victoria Hoelzer-Maddox, and Erin Shi for their assistance.

References

Alrababah, M. A., & Alhamad, M. N. (2006). Land use/cover classification of arid and semi-arid Mediterranean landscapes using Landsat ETM. *International Journal of Remote Sensing, 27*, 2703–2718.

Chinese National Land Resource Survey Committee. (1984). *Chinese national technical standard for land use survey*. Beijing: China Agricultural Press (in Chinese).

Coblentz, D. D., & Riitters, K. H. (2004). Topographic controls on the regional-scale biodiversity of the south-western USA. *Journal of Biogeography, 31*, 1125–1138.

Dewidar, K. (2004). Detection of land use/land cover changes for the northern part of the Nile delta (Burullus region), Egypt. *International Journal of Remote Sensing, 25*(20), 4079–4089.

Jarvis, A., Reuter, H. I., Nelson, A., & Guevara, E. (2006). *Hole-filled SRTM for the globe Version 3*, Retrieved from the CGIAR-CSI SRTM 90 m Database: http://srtm.csi.cgiar.org.

Jarvis, A., Rubiano, J., Nelson, A., Farrow, A., & Mulligan M. (2004). *Practical use of SRTM data in the tropics – Comparisons with digital elevation models generated from cartographic data* (Working Document No. 198). Cali: International Centre for Tropical Agriculture (CIAT), 32.

Jiang Q. (2002). LUCC and accompanied soil degradation in China from 1960's to 1990's. *Journal of Geoscience Research of Northeast Asia, 5*(1), 62–71.

Keuchel, J., Naumann S., Heiler M., & Siegmund, A. (2003). Automatic land cover analysis for Tenerife by supervised classification using remotely sensed data. *Remote Sensing of Environment, 86*, 530–541.

Kuemmerle, T., Radeloff, V. C., Perzanowski, K., & Hostert P. (2006). Cross-border comparison of land cover and landscape pattern in Eastern Europe using a hybrid classification technique. *Remote Sensing of Environment, 103*, 449–464.

Liu, J. G., & Diamond, J. (2005). China's environment in a globalizing world. *Nature, 435*, 1179–1186.

Nguyen, M. Q., Atkinson, P. M., & Lewis, H. G. (2005). Superresolution mapping using a Hopfield neural network with LIDAR data. *IEEE Geoscience and Remote Sensing Letters, 2*, 366–370.

Southworth, J. (2004). An assessment of Landsat TM band 6 thermal data for analyzing land cover in tropical dry forest regions, *International Journal of Remote Sensing, 25*, 689–706.

State Forestry Administration (SFA). (2007). *The State Forestry statistics* (in Chinese).

Tang, J., Wand, L., & Zhang, S. (2005). Investigating landscape pattern and its dynamics in Daqing, China. *International Journal of Remote Sensing, 26*(11), 2259–2280.

Tong, C., Hall, C. A. S., & Wang, H. (2003). Land use change in rice, wheat and maize production in China (1961–1998). *Agriculture, Ecosystems & Environment, 95*(2–3), 523–536.

Turner II, B. L., Lambin, E. F., & Reenberg, A. (2007). The emergence of land change science for global environmental change and sustainability. *PNAS, 104*(52), 20666–20671.

Wang, Z., Zhang, B., Zhang, S., Li, X., Liu, D., Song, K., et al. (2005). Changes of land use and of ecosystem service values in Sanjiang plain, northeast China. *Environmental Monitoring and Assessment, 112*, 69–91.

Xiao, J., Shen, Y., Ge, J., Tateishi, R., Tang, C., Liang, Y., et al. (2006). Evaluating urban expansion and land use change in Shijiazhuang China, by GIS and remote sensing. *Landscape and urban planning, 75*, 69–80.

Xu, J. T., Yin, R. S., Li, Z., & Liu, C. (2006) China's ecological rehabilitation: progress and challenges. *Ecological Economics 57*(4), 595–607.

Xu, C., Liu, M., An, S., Chen, J. M., & Yan, P. (2007). Assessing the impact of urbanization on regional net primary productivity in Jiangyin county, China. *Journal of environmental management, 85*, 597–606.

Yin, R. S., Xu, J. T., Li, Z., & Liu, C. (2005). China's ecological rehabilitation: unprecedented efforts, dramatic impacts, and requisite policies. *China Environment Series 6*, 17–32.

Yuan, F., Sawaya, K., Loeffelholz, B., & Bauer, M., (2005). Land cover classification and change analysis of the Twin Cities (Minnesota) Metropolitan area by multitemporal Landsat remote sensing. *Remote Sensing of Environment, 98*, 317–328.

Zha, Y., Liu, Y., & Deng, X. (2008). A landscape approach to quantifying land cover changes in Yulin, Northwest China, *Environment Monitor Assessment, 138*, 139–147.

Chapter 5
Modeling the Driving Forces of the Land Use and Land Cover Changes Along the Upper Yangtze River

Qing Xiang, Runsheng Yin, Jiutao Xu, and Xiangzheng Deng

Abstract Induced by high population density, rapid but uneven economic growth, and long-time resource exploitation, China's upper Yangtze basin has witnessed remarkable changes in land uses and covers, which have resulted in severe environmental consequences, such as flooding, soil erosion, and habitat loss. This paper examines the causes of the land use and land cover changes (LUCC) along the Jinsha River, one primary section of the upper Yangtze, aiming to better understand the human impact on the dynamic LUCC process and to provide necessary policy actions for sustainable land use and environmental protection. Using a panel dataset covering 31 counties over four time periods from 1975 to 2000, the study develops a fractional logit model to empirically determine the effects of socioeconomic and institutional factors on changes for cropland, forestland, and grassland. It is shown that population expansion, food self-sufficiency, and better market access drove cropland expansion, while industrial development contributed significantly to the increase of forestland and the decrease of other land uses. Similarly, stable tenure had a positive effect on forest protection. Moreover, past land use decisions were less significantly influenced by the distorted market signals. The policy implications of these findings and future directions of research are also discussed.

Keywords Land use and land cover changes · Driving forces · Fractional logit model · Upper Yangtze basin · Policy action

5.1 Introduction

A better understanding of the causes and consequences of land use and land cover changes (LUCC) is essential for global change studies because of their tremendous effects on carbon and water cycles, ecosystem functions, and human

Q. Xiang (✉)
Department of Forestry, Michigan State University, East Lansing, MI 48824, USA
e-mail: xiangqin@msu.edu

R. Yin (ed.), *An Integrated Assessment of China's Ecological Restoration Programs*,
DOI 10.1007/978-90-481-2655-2_5, © Springer Science+Business Media B.V. 2009

welfare (Turner, Meyer, & Skole, 1994; Geoghegan et al., 2001; Müller & Zeller, 2002; USGCRP, 2004). With continuous population growth and rapid economy development, China has witnessed substantial changes in land uses over the past several decades, and these changes have resulted in severe environmental consequences, such as flooding, soil erosion, and habitat loss. All of these have led to serious concerns about the sustainability of China's development (Liu, Liu, Zhuang, Zhang, & Deng, 2003). Thus, how to allocate the limited land resources so as to simultaneously satisfy the demands for food, natural resource materials, urban expansion, and quality environment has become a great challenge.

The upper Yangtze basin in China constitutes a great site for the LUCC research. The Yangtze River is the country's longest and the world's third largest river, which starts out from the Tibetan Plateau in the west, courses 6,300 km across 11 provinces (autonomous regions), and finally flows into the East China Sea. The Yangtze basin, with a total area of 180 million ha (19% of the country's area), nurtures around 40% of China's population, possesses 40% of the country's potential hydro-power, and contributes to more than 40% of China's GDP (Du, 2001). However, the development of the whole basin has been threatened by the environmental deterioration in the upper reaches.

The upper reaches of the Yangtze River refer to the vast area west of Yichang, Hubei, with a total area of over 105 million ha. The region is known for its rich biodiversity and complex geography. It features a wide variety of ecosystems that have been recognized as a major biodiversity hotspot (Conservation International, 2002). Accompanying the diverse ecosystems are the extreme fluctuations in topography and landscapes including high mountains and deep gorges in the west to hills and lowlands in the east. Such sharp variations make the region vulnerable, but it is the malpractices of human land use that worsen the situation. Deforestation, farmland expansion, and grassland degradation have seriously damaged native vegetation covers, causing severe environmental problems (Du, 2001; Xu, Katsigris, & White, 2002; Loucks et al., 2001). The deteriorated environment not only reduces land productivity in the region and threatens the lifespan and effectiveness of the Three-Gorges Dam, but also imposes large risks on economic development and people's livelihoods in the middle and lower reaches of the Yangtze. Therefore, it is important and imperative to understand how the regional LUCC are affected by various factors, so that appropriate policy adjustments can be made for more sustainable land use.

However, few rigorous studies have been done on the LUCC driving forces in the upper Yangtze. The existing works are mostly concerned with the long-term food security because of arable land loss that is undermining China's food production capacities. These works either examined cropland changes under different socioeconomic scenarios at the national level (Fischer & Sun, 2001; Verburg, Veldkamp, Espaldon, & Mastura, 2002; Zhang, Mount, & Boisvert, 2003), or analyzed regional arable land losses induced by urbanization and infrastructure or industrial expansion especially in the metropolitan areas (Yeh & Li, 1998; Ji et al., 2001), or looked into the effects of cropland suitability shifts on food production (Li, Peterson, Liu, & Qian, 2001). As stated by Liu et al. (2003), "... In the future, we need to study thoroughly the impact of human social and economic activities on land-use change at regional scales (p. 384)..."

This paper aims to gain a better understanding of the LUCC process in the upper Yangtze region and thus provide the essential knowledge for taking appropriate policy actions in achieving more sustainable land use. Specifically, we will develop and estimate a sound econometric model to determine the driving forces for changes in cropland, forestland, and grassland in the region; and we will do so by incorporating various socioeconomic and institutional as well as biophysical factors in a spatially explicit way. Our results show the driving forces have different effects on different land uses and the economic growth and institutional change have played important roles in affecting the LUCC. It is expected that the study will provide significant insights concerning the regional land policy, resource management, and environmental protection. Of course, these insights also are of relevance to other regions in China and indeed other developing countries. The paper is organized as follows. A brief description of the study site will be given in the next section, followed by the method and data sections. Then, estimation results will be presented before concluding remarks.

5.2 Study Site

Because of the expansiveness of the upper Yangtze basin, the study site was selected along the Jinsha River, part of the upper Yangtze with a length of 2,290 km. The Jinsha River refers to the section starting from Yushu County in Qinghai province, flowing across Qinghai, Tibet, Yunnan, and Sichuan, and ending in Yibin of Sichuan. The total area of this catchment is about 34 million ha, but included for this study are 31 counties fully located inside it (97.7°–104.8°E, 25.4°–32.7°N) with 14 million ha. Nineteen counties are in Yunnan, with an area of 7.2 million ha; the other 12 in Sichuan have an area of 6.8 million ha (Fig. 5.1).

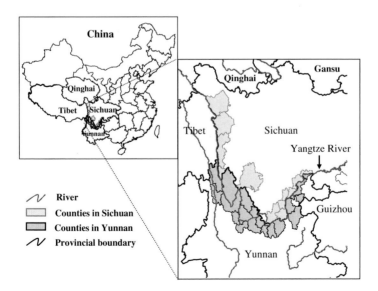

Fig. 5.1 The Study Region in Sichuan and Yunnan of China

The Jinsha River basin is known for its sharp descent, fragile geology, and severe soil erosion. The main stream precipitates by 3,280 m, while the elevation of the basin ranges from 300 to 6,140 m. Among the 31 counties, six have at least 70% of their lands at altitudes higher than 3,000 m, and 21 have at least 50% of their lands at altitudes between 1,500 and 3,000 m. Meanwhile, lands in 13 counties have slopes up to 60°. The geological structure of the lower Jinsha basin is dominated by Triassic shale and sandstone, which weather rapidly in the subtropical monsoon climate and yield soils that are susceptible to erosion (Lu, 2005). The steep slopes and fragile soils make the Jinsha basin a major sediment source to the Yangtze. It is estimated that annual soil loss from the upper Yangtze region averages 1.57 billion ton, accounting for 71.4% of the total soil loss of the whole Yangtze basin. About 45% of soil loss of the upper Yangtze region comes from the Jinsha basin (Pan, 1999).

Unfavorable natural conditions and poor transportation infrastructure make the region relatively isolated from the outside and suffer from a high incidence of poverty. In 2002, for example, the GDP of Ganzi Tibetan prefecture was ranked second to the last in Sichuan and even worse, the net per capita income of rural households (RMB 900/y) ranked the last in the province – less than half of the provincial average (Sichuan Statistics Yearbook, 2003)[1]. Of the 19 counties in Yunnan, 15 are national poverty counties (Wang, 2003), and 31.5% of the rural households had a per capita income lower than RMB 1,000/y (Yunnan Statistics Yearbook, 2003). Nonetheless, farming has long been a major income source in the region. From 1975 to 2000, the average value of annual farming output accounted for at least 55% of the total agricultural output (in addition to farming, official statistics for agriculture include animal husbandry and forestry). However, animal husbandry grew more rapidly than farming and forest sectors: its output share increased from 18% in 1975 to 35% in 2000. The output value from forestry was comparatively small, amounting to about eight percent of the total value of agricultural output, albeit the higher share of forestland.

5.3 Methods

5.3.1 Conceptual Model

Models used to examine the LUCC drivers are mostly derived from the land-rent maximization theory. The land owner is assumed to allocate a parcel of land of quality j at time t to the use that can generate the highest present value of a stream of net returns over time. For example, the net return for a parcel of cropland of certain quality at time t, W_t^a, equals the net present value of a stream of crop revenues minus cultivation costs, plus the present value of the land salvage value (Ahn, Plantinga, & Alig, 2000). The net return for a tract of forestland of certain quality at time t, W_t^f, is the net present value of a series of timber harvest rotations, determined by stumpage

---

[1]The Chinese government initially defined the poverty line as per capita income below 200 yuan in 1985. Based on inflation and other considerations, the figure has been adjusted upwards over time, reaching 1067 yuan in 2007 (China State Statistics Bureau 2008). A national poverty county is declared if a majority, but not necessarily all, of the local population lives below the poverty line.

prices and yields by species (Munroe & York, 2003). The land will be allocated to forestry if $W_t^f \geq W_t^a$, otherwise to farming (Plantinga, 1996).

The observed land use shares at the county level are derived by aggregating the land use choices of individuals who attempt to maximize land rents. For each land quality type j across a county, the land is allocated to different uses to maximize the total returns, subject to its availability (Miller & Plantinga, 1999; Ahn et al., 2000). That is:

$$Max \sum_k W_{jkt} \left(X_{kt}, h_{jkt} \right)$$

subject to:

$$\sum_k h_{jkt} = H_{jt} \qquad (5.1)$$

where k is the land use choice set, W_{jkt} is the net return function for the kth land use in quality class j at time t, h_{jkt} is the land area for the kth use in quality class j at time t, H_{jt} is the total land area with quality j at time t, and X_{kt} encompasses all the variables that affect the net return to land use k at time t. The solution to the problem is the optimal allocation $h_{jkt}{}^*$, expressed as:

$$h_{jkt}^* = h_{jkt}^* \left(H_{jt}, X_t \right) \qquad (5.2)$$

If the total land area for a county is $H(t) = \sum_j H_j(t)$, the sum of lands with different quality j, then the optimal share of land use k for this county at time t, p_k, is defined as:

$$p_k^* (X, t) = \frac{1}{H(t)} \sum_j h_{jkt}^*(H_{jt}, X_t) \qquad (5.3)$$

Notably, while a parcel of land with a certain quality can be allocated to different uses, the share of each land use differs across lands with different qualities or land characteristics. What we are interested in here is not only how the land is allocated to different uses at a given time, but also how changes happen to each use over time.

5.3.2 Estimation Method

Following the theoretical model for land allocation, the share of a land use can be defined as (Miller & Plantinga, 1999):

$$y_{it}^k = p_{it}^k (X_{it}) + \varepsilon_{it}^k \qquad (5.4)$$

where y_{it}^k and p_{it}^k are the observed and expected shares of land use k in county i at time t, and ε_{it}^k is the independently and identically distributed error term. The sum of y_{it}^k or p_{it}^k equals to unity when the range of k covers each land use in county i at time t.

It should be pointed out that y and p are bounded between zero and one, implying that it is not appropriate to express y or p as a linear function of the explanatory variables and thus to estimate it with a conventional method. A potential problem is that the fitted value of y or p may fall outside the unit interval. To avoid this problem, p can be modeled as a logistic function:

$$E\left(y_{it}^k \mid X_{it}\right) = p_{it}^k = \frac{\exp\left(X_{it}\,\beta_k\right)}{1 + \exp\left(X_{it}\,\beta_k\right)} \tag{5.5}$$

and then the observed land share y for use k is expressed as:

$$y_{it}^k = \frac{\exp\left(X_{it}\,\beta_k\right)}{1 + \exp\left(X_{it}\,\beta_k\right)} + \varepsilon_{it}^k \tag{5.6}$$

where β is the coefficient vector. The above specification, called the fractional logit model, ensures that predicted values for y ranges between zero and one. It is assumed that ε satisfies a logistic distribution. Note that while β_n gives the sign of the partial effect of the nth explanatory variable on a land use, its magnitude cannot represent the partial effect of that explanatory variable on the dependent variable (Wooldridge, 2002).

A popular approach to coefficient estimation is to transform the above model so that the log-odds of the dependent variable have a conditional expectation on the linear form of explanatory variables: $E\left(\log[y/(1-y)]/X\right) = X\beta$, which is estimated by the *OLS* method. But such a transformation has certain drawbacks. First, it cannot be used directly if the dependent variable takes on boundary values, zero and one. Second, it is difficult to interpret the coefficients because without further assumptions it is impossible to recover how y is expressed by explanatory variables (Wooldridge, 2002). To address these problems, the fractional logit model is estimated with the quasi-Maximum Likelihood Estimator (MLE), which provides a consistent estimate of β when $E(y|x)$ is expressed as a logistic form. Meanwhile, the potential problems of heteroskedasticity and serial correlation in variance can be taken care of with the econometrics software.

In principle, the fractional logit Equation (5.6) needs to be established for each type of land use k, and the sum of dependent variables equals to one. To ensure the identification of these equations, only k-1 equations are estimated. Because the total land area is classified into four primary categories in this study, a system of three equations will be estimated. The effects of explanatory variables on the fourth land use type equal to unity minus the sum of the effects on the other three.

5.4 Data Description

The dataset, covering the 31 counties over five time points (mid-1970s, mid-1980s, late 1980s, mid-1990s, and late 1990s), is used in the fractional logit model to elucidate the LUCC drivers from 1975 to 2000. The dataset consists of two parts: the dependent variables – land use shares derived from satellite images, and the explanatory variables – biophysical and socioeconomic factors gathered from multiple sources.

5.4.1 Land Use Data

The land use data are derived from Landsat Multi-Spectral Scanner (MSS), Thematic Mapper (TM), and Enhanced Thematic Mapper (ETM+) images. In cases where certain images are missing or of poor quality, those from adjacent years are taken to obtain the information for a given time point. The land use/cover data of the mid-1970s (hereafter referred to as "1975 data") are derived from MSS images of 1973–1977, late 1980s data ("1990 data") are derived from 1988 and 1989 TM images, mid-1990s data ("1995 data") are derived from TM images of the year 1995, and late 1990s data ("2000 data") are derived from TM/ETM+ images for 1999 and 2000.

The data for 1990, 1995, and 2000 are a subset of China's national land cover dataset created by the Chinese Academy of Sciences (CAS). Due to the unavailability of images for the mid-1980s, a land cover map for 1985 was scanned and digitalized to generate the needed data. The classified land use/cover data are re-sampled to form a raster-format dataset with a resolution of 100 m, and then overlapped with a county boundary map to generate the corresponding county-level data (Table 5.1). Tests by CAS (Liu et al., 2003) show that the accuracy rate for 1975 data is 88%; the accuracy rate for 1990, 1995, and 2000 is 92.92, 98.40, and 97.45%, respectively.

Table 5.1 Land Use Patterns in the Study Region from 1975 to 2000

Land uses/ covers (ha)	1975	1985	1990	1995	2000	Changes between 1975 and 2000
Cropland	1,603,687	1,614,621	1,561,700	1,563,426	1,560,719	–42,968
Forests	7,487,370	7,530,783	7,498,984	7,551,545	7,466,451	–20,919
Grassland	4,242,754	4,117,841	4,310,880	4,224,982	4,388,228	145,474
Others	707,821	778,387	670,068	701,679	626,234	–81,587

Note: Data are from the Institute of Geographic Science and Natural Resources, Chinese Academy of Science (Liu et al., 2003).

5.4.2 Explanatory Variables

We use the procurement prices for grain, log, and livestock to represent the economic returns from cropland, forestland, and grassland. Relative to these market signals, however, decisions on land use in China were and still are influenced by government regulation and population pressure. For cropland, farmers sign contracts with local governments to manage it, and contracts are valid for decades and seldom adjusted so as to provide stable "use rights." In principle, cropland expansion on grassland or forestland is prohibited; but in practice, such encroachment does happen due to the necessity for food production and the difficulty in regulation enforcement. Because of uncertain use rights, however, reclaiming cropland could be induced by contemporary price change, but not by the long-term profitability. Grain procurement prices faced by producers were controlled by the government, and their effects on land use decisions might not be as apparent as expected because they were depressed. In forestry, timber harvest and distribution were largely controlled by the government, and the state-owned enterprises and collective forest entities had little motivation for long-term management. It is thus reasonable to assume that forestland changes were not greatly affected by the limited price movements. Similarly, the modest livestock price changes might not induce significant grassland shifts. Nevertheless, output prices can have significant cross effects.

Costs are not easily identifiable for crop, timber, and livestock production at the county level. As an alternative, we adopt the approach used by Chomitz and Gray (1996) and elaborated by Kaimowitz & Angelsen (1998), who observed that the distance of a parcel of land to roads, reflecting market access, affects both output and input costs and thus land use patterns. The road length in a county is used to capture the transportation cost and market access for that county. And we expect that compared to grassland and forestland, cropland is more likely allocated close to settlement centers.

Industrial development is included as an explanatory variable as well. This is because while industrial development may take away some fertile cropland, it promotes transfer of surplus rural labor to off-farm activities, which reduce pressures on natural resources and help environmental conservation. Off-farm activities also alter the opportunity cost for rural labor, which constrains labor available to extensive farming (McCracken et al., 1999) and improves farmers' income and abilities to adopt new technologies. As a result, enhanced land use productivity can better satisfy livelihood needs and therefore reduce resource overexploitation.

Population expansion is widely used as a determinant of land use changes (Mertens, Sunderlin, Ndoye, & Lambin, 2000), and its main effect is to cause cropland encroachment on forestland and grassland and related resource degradation (Yin & Li, 2001). Soil characteristics influence land allocation by determining land suitability for different land uses and productivity. But measuring soil characteristics for a county is hard because of the large variations of the soil features. So, the average elevation of a county is used to represent soil features as well as temperature and other biophysical conditions that affect land use[2].

The food self-sufficiency policy and forest tenure arrangement are two major political-institutional factors that have affected land use patterns in the study region. The former, reflected in grain procurement quota, encouraged cropland expansion on slopes previously covered by forest or grassland (Xu, Yin, Li, & Liu, 2007). It is thus hypothesized that a decreasing quota, as a sign of relaxing the policy, should benefit the restoration of vegetation covers. The latter, if clearly defined and enforced, forms the basis for at least stable forest management. In the study area, around 30% of the forestland is owned by the state. This forestland was seldom

---

[2]Elevation will not be listed in the summary statistics of variables because it is a time-constant variable. It ranges from 295 meters (m) to 6109 m for the study region, with a mean of 3070 m.

Table 5.2 Summary of Variables Used in the Fractional Logit Model

	Measurement	1975	1985	1990	1995	2000
Population	1,000 Persons	6,101	6,815	7,141	7,592	7,908
Population density	Person/km^2	0.66	0.73	0.78	0.82	0.86
Grain price[1]	Index (1990 = 1)	0.587	0.669	0.972	1.619	2.035
Log price[1]	Index (1990 = 1)	0.589	0.531	0.927	1.046	1.294
Livestock price[1]	Index (1990 = 1)	0.860	0.863	1.085	1.873	2.218
Industry output (IO)[2]	1,000 RMB	69,534	141,243	169,494	516,442	698,396
Agricultural output (AO)[2]	1,000 RMB	189,063	329,188	351,141	533,255	729,546
IO/AO		0.320	0.403	0.443	0.768	0.762
Highway	km	6,132	9,836	11,592	14,173	28,059
Highway rate	km/ha	0.0009	0.0014	0.0016	0.0021	0.0031
Grain quota[3]	Ton	170,423	141,292	111,231	115,691	107,481
Per capita quota	Ton/person	0.030	0.028	0.016	0.016	0.014
State forest share		0.338	0.331	0.322	0.319	0.314

Note:
[1] Grain, log, and livestock price indices are provincial aggregates.
[2] Agricultural and industry outputs are output values at 1990 constant price.
[3] Grain quota is from the local grain bureau, and the state forest share is from the local forest bureau. All other data come from local statistics bureaus.

converted to other uses due to its stable and clear tenure, although forest degradation happened because of over-harvesting. In contrast, for sloping lands that belonged to the collectives or those without clear ownership, forestland or grassland loss to cropland often occurred (Xu et al., 2007). Thus, the share of state-owned forest is employed to proximate forest tenure stability. It is not suggested, though, that state ownership is superior and does not need reform.

Table 5.2 summarizes all the variables. Note that a variable's value at a given point of time is the average value from the adjacent years whose range is the same as that of the land cover data. For instance, the land use/cover data in 1975 is derived from the remote sensing images of 1973–1977; so correspondingly, data for each explanatory variable in 1975 is the mean value of 1973–1977. Appendix details the definitions and variations of some of these variables.

5.5 Estimation Results

Including all the explanatory variables, our empirical model becomes:

$$Y_{it} = f\left(GP_{it}, FP_{it}, LP_{it}, IND_{it}, POP_{it}, ROAD_{it}, GQ_{it}, SF_{it}, E_{it}\right) + \varepsilon_{it}$$

where i denotes county and t denotes time. Y represents cropland share, forest share, or grassland share; GP is the price index for grain, FP the price index for logs, *LP* the price index for livestock; *IND* is the ratio of industry output to gross agricultural output; *POP* denotes population density, *ROAD* denotes the highway rate, *GQ* represents per capita grain quota, *SF* represents the share of state-owned forests, and *E* denotes the elevation. Province and year dummies are also added to control variations across provinces and over time. The model is estimated with STATA software. Endogeneity test indicates that all explanatory variables can be taken as exogenous ones.

Table 5.3 lists the estimated results for the shares of cropland, forestland, and grassland. To save space, coefficients for province and year dummy variables are omitted. It should be noted that the listed coefficients are the corresponding elasticities calculated according to the land use changes, but not β. This way of presentation can indicate the extent of driving forces' impacts on each type of land use. In general, signs of most coefficients are as expected and statistically significant, and the effects of explanatory variables are different on different types of land use.

Two price coefficients are statistically significant. First, grain price has a significant negative effect on the change of forestland share; one percent increase of grain prices can reduce the forestland share by 0.37%, holding other variables constant. This indicates that farmers indeed seek short-term farming profits from the increased grain prices by encroaching forestland. Second, the effect of livestock price on cropland share is positively significant at 10% level; one percent increase of livestock price can result in 0.94% increase of the cropland share when controlling other variables. This suggests that some crops are planted for feeding domestic animals, so a higher livestock price drives more feedstock production. The insignificance of most

Table 5.3 Estimation Results of Fractional Logit Model for Primary Land Use

Explanatory variables	Cropland	Forest	Grassland	Others
Grain price	− 0.519 (0.358)	− 0.367 (0.180)**	0.357 (0.342)	− 0.428
Log price	0.011 (0.101)	− 0.087 (0.058)	0.067 (0.089)	− 0.215
Livestock price	0.944 (0.565)*	0.192 (0.368)	0.087 (0.694)	− 4.251
Industry/ agricultural output	− 0.047 (0.011)***	0.068 (0.027)**	− 0.114 (0.059)*	0.919
Highway rate	0.215 (0.077)***	0.018 (0.054)	0.105 (0.140)	− 1.721
Population density	0.480 (0.028)***	− 0.335 (0.082)***	0.157 (0.134)	−0.732
Per Capita Grain quota	0.037 (0.020)*	− 0.005 (0.011)	− 0.015 (0.034)	0.108
Share of state-owned forest	− 0.132 (0.085)	0.116 (0.072)*	− 0.133 (0.107)	1.023
Elevation	− 1.232 (0.128)***	− 0.633 (0.139)***	1.222 (0.335)***	− 6.509

Note:
1. Numbers in parentheses are standard error of the coefficient "*", "**" and "***" represent 10, 5, and 1% significance level, respectively.
2. There were 122 observations used in the estimation, because of some missing values for high way variable. So, the degree of freedom was 107.
3. Result for other lands are derived, thus without significance level.

other price variables partially proves our prior knowledge that the LUCC might not be driven much by price signals, especially for forestland and grassland. On the other hand, the use of a provincial index could have concealed the price variations across counties.

Industrial development significantly reduces the pressures on resource exploitation. The higher the industry output, the lower the shares of cropland and grassland, and the higher the share forestland. A one percent increase of industrial development can result in 0.05% decrease in cropland, 0.11% decrease in grassland, and 0.07% increases in forestland. As discussed earlier, economic development does affect the LUCC by altering land economic returns and labor opportunity costs. Consistent with our assumption, highway construction has significant effects on cropland changes. Holding other variables constant, one percent increase of the highway rate can result in a 0.22% increase in the cropland share. Insignificant effects of highway rate on forestland and grassland changes imply that given other conditions, forestland or grassland share in a county is not closely connected to its highway length. So, road construction has more influence on the changes of cropland use than anything else.

Demography significantly affects land use changes. Counties with higher population density observed higher cropland share and lower forestland share. Increased population resulted in cropland expansion to meet the growing food demand and imposed pressure on forest resources. One percent increase in population density causes the cropland share to grow by 0.48%, while it induces forestland to decline by 0.33%. Although the effect of population growth on grassland changes is statistically insignificant, the positive sign makes sense. Results from the three equations show that the share for other lands declined with the population growth, which suggests that farming encroaches on unused lands or even bodies of water because of limited cropland resource.

Grain procurement quota has significant effects on cropland expansion. One percent increase of per capita grain quota could induce a 0.04% increase of the cropland share. No significant effects of grain quota on forestland and grassland changes are found, but the signs of coefficients are as expected: an increase of grain quota is associated with a decrease in forestland and grassland. So, it can be inferred that eliminating grain quotas will reduce not only burdens on farmers but also cropland expansion at the expense of forestland or grassland. Tenure indeed has a significant influence on forest resource status. The share of forestland increases by 0.12% when the state-ownership is one percent higher, holding other variables constant. Additionally, the share of other lands goes up with the increase of state-owned forestland. Finally, the estimation also proves that land use allocation varies significantly with altitude. The cropland and forestland shares decrease with altitude, but more grassland is found in the high elevation.

5.6 Conclusions and Discussion

This study develops and estimates a fractional logit model to examine the driving forces of the LUCC in the upper Yangtze basin. Results indicate that industrial development had a significant effect on reducing cropland expansion and conserving forest resources, whereas population growth put pressures on land resources and contributed to deforestation and grassland degradation. In the past, land use decisions were made to capture immediate profits, but they were not significantly influenced by the long-term price signals. In addition, institutional and policy factors played critical roles in shaping the land use patterns: lowering grain quota levied on farmers reduced cropland expansion, and stable forest tenure led to a higher share of forestland.

These and other results carry significant policy implications. First, off-farming opportunities not only increase farmers' income, but also lure them out of rural areas and thus reduce their disturbance to land covers. With the increasing population, development of non-farming sectors has become an important way to alleviate poverty as well as to protect natural resources. Local governments should provide job services to facilitate farmers' pursuit of off-farming activities. Second, market mechanism should be promoted in allocating land resources. That prices did not significantly influence land use decisions in the past is partially because these prices were not real market signals but government-controlled ones, and such distorted prices could not adequately guide long-term land use decisions. For instance, the growth rate of log procurement price was lower than that of livestock price and grain price, and the log price did not reflect higher market demand for forest products and discouraged reforestation and forest management. Reforestation and afforestation will improve if forest managers face higher and more transparent log prices and thus expect to get reasonable returns in the long run.

The food self-sufficiency policy was not conducive to efficient and sustainable land use because the grain procurement quota disrupted the trade and distribution of agricultural products across the nation, and caused more land and other inputs to be used in crop production. It is not necessary to meet food demand with local production for a region like our study site that possesses poor farming conditions and limited cropland. However, with abundant grassland and forestland, farmers should specialize in livestock and forest industries and establish their comparative advantage in the marketplace.

Moreover, clearly defined tenure arrangement encourages long-term planning and protection of forest resources. State ownership represents a stable forest tenure, which has reduced the possibility of forest conversion. However, unclear beneficiaries of collective forests and distorted market prices discouraged farmers from forest management and investment, and consequently the collective forests were more likely to be degraded and even converted to other uses. Results of this study imply that it is important to implement tenure reforms for the collective forests, including the clarification of use and benefit rights, the creation of a well-functioning monitoring and enforcement system, and dissemination of transparent and fair market information to the local forest managers. In doing so, it is expected that the forest conversion to other uses will slow down or even reverse, forest investment will increase, and sustainable forest management will follow.

While the fractional logit model developed well explains the driving forces of the regional LUCC, more needs to be done in order to enhance our understanding of the complicated land use processes. The logit form of the share functions is ad hoc to some extent. As applied in this study, it does not treat relevant variables as endogenous or account for the feedback effects and thus is insufficient to illuminate the dynamic interactions among different variables. Therefore, future study should develop more sophisticated models that incorporate the interactive processes and environmental consequences into the analysis of the LUCC driving forces (Turner, Lambin, & Reenberg, 2007). Such efforts will shed new light on the fundamental question of how the land use changes have happened and help generate more reliable projections on the future LUCC. Moreover, the effect of technological progress on the LUCC should be considered. Technology progress is a critical solution to sustaining the livelihood of an expanding population on a limited land base (Müller & Zeller, 2002). But technology progress itself is determined by socioeconomic changes and resource endowment (Ruttan, 2001). Thus, incorporating endogenous technology progress into the analytic framework in examining the LUCC driving forces more effectively will be a major research step, which can lend great insights to the quest for sustainable development. Undoubtedly, these tasks will be accomplished only if more comprehensive datasets, particularly those longitudinal ones with long time series and spatially explicit observations, can be built.

Appendix: Definition and Variation of Some Key Variables

Land use is categorized as cropland, forestland, grassland, and other lands. Forestland, grassland, and cropland are major land-use types for the study region. In 2000, these three land-use types accounted for 53, 30, and 11%, respectively. All other lands accounted for about five percent of the total land (Table 5.1). Compared to 1975, areas of cropland, forest, and other lands in 2000 decreased, while grassland increased. Notably, for each of four periods from 1975 to 2000, the percentage change of each land use was not large. Because the study area has 14 million ha, a change even as small as one percent was actually not of small magnitude. On the other hand, the original image processing at the national scale could have obscured the county-level LUCC. Moreover, opposite land conversions (e.g., from cropland to forest and vice versa) always take place, leading to the lack of relative variation for each land use over time.

The land conversion information is insightful for understanding the dynamic LUCC process. The extent of cropland conversion to forestland, grassland, and other lands was larger before 1990 than that thereafter. Between 1975 and 1990, about ten percent of cropland was converted out, whereas from 1990 to 2000 only four percent of cropland was converted out. Similarly, grassland conversion to cropland, forestland, and other lands was at a larger scale before 1990. Except for the period of 1975–1985 when a majority of grassland was converted to other lands, most of the grassland was converted to forestland. Meanwhile, grassland was also the major outlet of forest conversion. In contrast, conversion between grassland and forestland for each period made the total area for forestland and grassland look stable, which indicates the necessity to examine the causes of the changes for different land uses at a finer scale, such as the county level.

Prices used are the provincial indices, meaning that at each time point the value of each price is the same for all counties of the province. All prices increased continuously from 1975 to 2000. Grain and livestock prices rose at an annual rate of over 12% in the early 1990s, while the late 1980s witnessed the fastest log price increase. Growing livestock price over time partially explains why livestock husbandry is preferred as a means of improving rural income. Log procurement price was more stagnant compared to the other two. This is partly why the farmers have little incentive for long-run forest investment and management, and instead they harvest when the price goes up or when they have access to the forest to capture immediate profits.

Highway mileage in a county is used to represent road length. China's statistics provide a standard definition of highway, so there is no ambiguity or discrepancy for data of highway length across counties over time. Highway rate, the ratio of highway length to county area, is the variable used in the analysis to remove the effect of county size.

Industrial development is defined as the ratio of industry output to gross agricultural output. Despite the dominance of the agricultural sector in the economy, its share declined. In 1975, the ratio of industry output value to the agricultural output was 0.3, and it increased to 0.4 in 1990 and further soared to 0.76 in 2000. Compared to that in 1975, industry output in 1990 more than doubled, and in 2000 increased by nine times. This is because the annual growth rate of industry output was much higher than that of agriculture.

Population density, total population divided by county area, is used in our modeling. Total population kept increasing over time, but at a declining rate. The annual population growth rate was 1.3% between 1975 and 1985, then decreased to 0.83% in the late 1990s. Accordingly, the population density also rose over time at a declining rate.

Grain procurement quota includes a portion for paying agricultural tax and another portion that is mandated to be sold to the state-owned procurement bureau at lower prices. Grain procurement quota declined gradually, and in 2000 it decreased to 60% of the 1975 level. Because of such a large decline, grain quota per rural resident also decreased dramatically, from 30 kg in 1975 to around 14 kg in 2000. The decrease of grain quota over time implies that the food self-sufficiency requirement became gradually out of date as the agricultural produce market became more developed. Since 2003, China has terminated the quota-based agricultural tax nationwide.

Share of state-owned forests in Sichuan is around 50%, while in Yunnan province it is just 20%. The share of state-owned forest decreased slightly over time in both provinces, and there were two reasons for this. First, disputes about land ownership were gradually resolved between local government and communities, with the latter having taken up some forests from the former. Second, afforestation and conversion of sloping cropland to forests increased the total as well as collective forestland gradually, and thus the percentage of the state forest declined.

References

Ahn, S., Plantinga, A. J., & Alig, R. J. (2000). Predicting future forestland area: A comparison of econometric approaches. *Forest Sciences, 46*(3), 363–376.

China State Statistics Bureau. (2008). China Statistics Yearbook. Beijing: China Statistics Press.

Chomitz, K. M., & Gray, D. A. (1996). Roads, land use, and deforestation: A spatial model applied to Belize. *World Bank Economic Review, 10*(3), 487–512.

Conservation International. (2002). *Biodiversity hotspots*. Retrieved from http://www.biodiversity hotspots.org/xp/hotspots/China

Du, S. F. (2001). *Environmental economics*. Beijing: Encyclopedia Press.

Fischer, G., & Sun, L. X. (2001). Model based analysis of future land-use development in China. *Agriculture, Ecosystems and Environment, 85*, 163–176.

Geoghegan, J., Villar, S. C., Klepeis, P., Mendoza, P. M., Himmelberger, Y. O., Chowdhury, R. R., et al. (2001). Modeling tropical deforestation in the southern Yucatán Peninsular region: Comparing survey and satellite data. *Agriculture, Ecosystems and Environment, 85*, 25–46.

Ji, C. Y., Liu, Q., Sun, D., Wang, S., Lin, P., & Li, X. (2001). Monitoring urban expansion with remote sensing in China. *International Journal of Geographical Information System, 22*(8), 1441–1455.

Kaimowitz, D., & Angelsen, A. (1998). *Economic models of tropical deforestation: A review.* Bogor, Indonesia: Center for International Forestry Research.

Li, X., Peterson, J. A., Liu, G., & Qian, L. (2001). Assessing regional sustainability: The case of land use and land cover change in the Middle Yiluo Catchment of the Yellow River Basin, China. *Applied Geography, 21*, 87–106.

Liu, J. Y., Liu, M. L., Zhuang, D. F., Zhang, Z. X., & Deng, X. Z. (2003). Study on spatial pattern of land-use change in China during 1995–2000. *Science in China (Series D), 46*(4), 373–384.

Loucks, C. J., Lü, Z., Dinerstein, E., Wang, H., Olson, D. M., Zhu, C. Q., et al. (2001). Giant pandas in a changing landscape. *Science, 294*, 1465.

Lu, X. X. (2005). Spatial variability and temporal change of water discharge and sediment flux in the lower Jinsha tributary: impact of environmental changes. *River Research and Applications, 21*(2–3), 229–243.

McCracken, S. D., Brondizio, E. S., Nelson, D., Moran, E. F., Siqueira, A. D., & Rodriguez-Pedraza, C. (1999). Remote sensing and GIS at farm property level: Demography an defor-estation in the Brazilian Amazon. *Photogrammetric Engineering and Remote Sensing, 65*(11), 1311–1320.

Mertens, B., Sunderlin, W. D., Ndoye, O., & Lambin, E. F. (2000). Impact of macroeconomic change on deforestation in South Cameroon: Integration of household survey and remotely sensed data. *World Development, 28*(6), 983–999.

Miller, J. D., O., & Plantinga, J. A. (1999). Modeling land use decision with aggregated data. *American Journal of Agricultural Economics, 81*, 180–194.

Müller, D., & Zeller, M. (2002). Land use dynamics in the central highlands of Vietnam: A spatial model combining village survey data with satellite imagery interpretation. *Agricultural Economics, 27*, 333–354.

Munroe, D. K., & York, A. M. (2003). Jobs, houses and trees: Changing regional structure, local land-use patterns, and forest cover in Southern Indiana. *Growth and Change, 34*(3), 299–320.

Pan, J. G. (1999). The characteristics of water runoff and suspended sediment along the Jinsha River. *Journal of Sediment Research, 2*, 19–24.

Plantinga, A. J. (1996). The effects of agricultural policies on land use and environmental quality. *American Journal of Agricultural Economics, 78*, 1082–1091.

Ruttan, V. W. (2001). *Technology, growth, and development: An induced innovation perspective.* Oxford, UK: Oxford University Press.

Sichuan Statistics Yearbook. (2003). Chengdu, China: Sichuan Statistics Press.

Turner II, B. L., Lambin, E. F., & Reenberg, A. (2007). The emergence of land change science for global environmental change and sustainability. *PNAS, 104*(52), 20666–20671.

Turner II, B. L., Meyer, B. W., & Skole, D. L. (1994). Global land use/land cover change: towards an integrated study. *AMBIO, 23*(1), 91–94.

USGCRP. (2004). Land use/land cover change: USGCRP program element. Retrieved from http://www.usgcrp.gov/usgcrp/ProgramElements/land.htm.

Verburg, P. H., Veldkamp, W. S. A., Espaldon, R. L. V., & Mastura, S. S. A. (2002). Modeling the spatial dynamics of regional land use: the CLUE-S model. *Environmental Management, 30*(3), 391–405.

Wang, X. T. (2003). *Building an ecological shield along the Upper Yangtze river: Priorities and measures.* Beijing: China Agriculture Publishing House.

Wooldridge, J. M. (2002). *Econometric analysis of cross section and panel data.* Cambridge, MA: The MIT Press.

Xu, J. T., Katsigris, E., & White, A. (Eds.). (2002). *Implementing the natural forest protection program and the sloping land conversion program: Lessons and policy recommendations.* Beijing, China: China Forestry Press.

Xu, J., Yin, R. S., Li, Z., & Liu, C. (2007). China's ecological rehabilitation: Progress and challenges. *Ecological Economics, 57*(4), 595–607.

Yeh, A. G., & Li, X. (1998). Suitable land development model for rapid growth areas Using GIS. *International Journal of Geographical Information System, 12*(2), 169–189.

Yin, H., & Li, C. (2001). Human impact on floods and flood disasters on the Yangtze River, *Geomorphology, 41*(2–3), 105–109.

Yunnan Statistics Yearbook. (2003). Kunming, China: Yunnan Statistics Press.

Zhang, X., Mount, T. D., & Boisvert, R. N. (2003). *Industrialization, urbanization and land use in China.* IFPRI, Environment and Production Technology Division (Discussion Paper No. 58).

Chapter 6
An Integrative Approach to Modeling Land Use Changes: The Multiple Facets of Agriculture in the Upper Yangtze Basin

Runsheng Yin and Qing Xiang

Abstract Land change science has emerged as a fundamental component of global environmental change and sustainability research. Still, much remains to be learned before scientists can fully assess future roles of land use/cover changes (LUCC) in the functioning of the earth system and identifying conditions for sustainable land use. The objective of this chapter is to gain a better understanding of the complex interactions of human and natural drivers underlying LUCC. We do so by developing and estimating a novel structural model of land use and by using spatially explicit longitudinal observations from the upper Yangtze basin of China. Our analysis focuses on the multiple dimensions of agriculture—not only cropland use itself, but also grain production, soil erosion, and related technical change; and our data cover 31 counties over four time periods from 1975 to 2000. Our results show that technical change plays an important role in supplying food on a limited cropland; limiting cropland expansion in turn reduces soil erosion, which then benefits grain production in the longer term. It is also found that policies and institutions have significant impacts on land use and the status of soil erosion. Together, these results carry some great implications to sustainable land use and ecosystem management.

Keywords Land use and land cover change · Coupled human and natural processes · Driving forces · Structural model of multiple equations · Upper Yangtze basin

6.1 Introduction

Human-driven changes in the terrestrial surface of the planet hold broad significance for the structure and functions of ecosystems, with equally far-reaching consequences for human well-being (Turner et al., 2007). Past land-use and

R. Yin (✉)
Department of Forestry, Michigan State University, East Lansing, MI 48824, USA;
Ecosystem Policy Institute of China, Chaoyang District, Beijing 100102, P.R. China
e-mail: yinr@msu.edu

R. Yin (ed.), *An Integrated Assessment of China's Ecological Restoration Programs*,
DOI 10.1007/978-90-481-2655-2_6, © Springer Science+Business Media B.V. 2009

land-cover changes (LUCC), while enabling humans to appropriate an increasing share of the earth's resources, have profoundly altered its conditions and adversely affected carbon and water cycles and ecosystem functions (Foley et al., 2005; Geoghegan et al., 2001); and LUCC in the future will further intensify the trends of climate change, groundwater depletion, species extinction, and soil nutrient losses (Millennium Ecosystem Assessment, 2005). Hence, LUCC have been widely recognized as an important field of scientific research. The US Global Change Research Program stated that: "A better understanding of the processes, rates, causes, and consequences of land use change and land management practices is essential for many areas of global change research" (USGCRP, 2004). More recently, "land change science has emerged as a fundamental component of global environmental change and sustainability research" (Turner et al., 2007).

Significant progress in quantifying LUCC in the past decade notwithstanding, a lot remains to be learned before scientists can fully assess the future roles of LUCC in the functioning of the earth system and identifying conditions for sustainable land use (USGCRP, 2004, Lambin, Geist, & Lepers, 2003). Early studies of the LUCC causes tend to feature some sort of discrete choice or reduced-form model, which prescribes that the share or quantity of LUCC is determined by land rent, land and/or landowner characteristics, and other external factors (Walker & Moran, 2000; Kaimowitz & Angelsen, 1998; Chomitz & Gray, 1996). Among other things, these studies identify economic return, market access, industrial development, demographic change, and macro policy as major drivers of land use decisions. However, they have given little attention to the interactions of biophysical factors, such as soil and water conditions, and socioeconomic factors, such as technical change in the LUCC process (Veldkamp & Fresco, 1996; Lambin, Rounsevell, & Geist, 2000).

The study by Xiang et al. (Chapter 5) was conducted to illustrate the strengths and weaknesses of the current analytic approach. Using a fractional logit model, they examined the LUCC driving forces in China's Upper Yangtze basin. Their results show that industrial development had a significant effect on reducing cropland expansion and conserving forests, whereas population pressures contributed to deforestation and grassland degradation. Further, land use decisions were not significantly influenced by the distorted, and often depressed, price signals. In addition, institutional and policy factors played critical roles in shaping the land use patterns. Nonetheless, that type of model failed to capture the connectivity of various factors in influencing LUCC. Also, technical change was not incorporated. So, they came to the conclusion that more sophisticated modeling strategy is called for to reflect the dynamic LUCC linkages.

In fact, as a coupled natural-human process, not only are LUCC affected by the interactions of biophysical and socioeconomic factors, these factors also can in turn be affected by LUCC and their induced feedbacks. So far, few efforts have been made to investigate these complex relationships. Also, empirical linkages between proposed causal variables and LUCC commonly involve the more proximate factors to the land-use end of explanatory connections, such as subsistence farmers and deforestation; the root causes that shape the proximate ones, such as poverty or policy, tend to be difficult to connect to land outcomes, due to the number and complexity of the linkages involved (Turner et al., 2007). For instance, it is poorly understood how technical change can mitigate the pressures of population growth and ecological degradation (Müller & Zeller, 2002). To overcome these difficulties, it is essential that LUCC studies incorporate the relevant variables and account for their endogenous effects (Irwin & Geoghegan, 2001). To that end, however, continuous and long data series are needed.

The objective of this chapter is to address the above challenges in LUCC research. We will do so by developing and estimating a novel structural model of land use and by using spatially explicit longitudinal observations from the upper Yangtze basin of China. Our model will focus on the multiple dimensions of agriculture—not only cropland use itself, but also grain production, soil erosion, and related technical changes. That is, we will develop a system of four equations to characterize the various facets of agriculture and their interactions in elucidating the LUCC driving forces. Notably, cropland is a main category of land use in the study region, and data for an integrative study of the agricultural sector are more likely available.

It is expected that this effort will contribute to a better understanding of the complex human and nature processes underlying the LUCC. In particular, our results show that technical change plays an important role in producing food on a limited land base; limiting cropland expansion in turn reduces soil erosion, which then benefits grain production in the longer term. Our results also highlight that policies and institutions have significant impacts on land use and the status of soil erosion. Together, these results carry some important implications to sustainable land use and ecosystem management. The chapter is organized as follows. We present our methods and data in the next section, followed by estimated results. The final section contains the conclusions and discussion.

6.2 Methods and Data

6.2.1 Conceptual Framework

To disentangle the complex interactions of LUCC drivers, we will concentrate on the multiple facets of cropland use. Our model of the causal relationships for cropland use is composed of four interactive components: grain production, farming technical change, and soil erosion, in addition to cropland itself. Grain production is determined by labor, land, capital, and other inputs with embedded technical changes, such as irrigation and fertilization. Technical changes are induced by the relative prices or opportunity costs of production factors (Ruttan, 2001). That is, farmers make their decisions of land use and technical adoption in response to the external economic conditions. When land rent becomes lower compared to capital or labor costs, land-extensive technology as well as production mode will be employed. When land becomes scarcer and thus rent increases, substitution of land with labor or capital will occur, and corresponding farming technology will be adopted (Müller & Zeller, 2002).

Technical adoption enhances grain productivity, which will in turn affect land use decisions. Changes in land use patterns can then alter the status of resource scarcity, which has implications to resource rent (price) and technology adoption. Of course, technical change must be brought to bear by institutional changes, such as improved land tenure and price liberalization (Lin, 1992; Yin & Hyde, 2000). Furthermore, not only does land use change interact with technical change, but its environmental consequence also generates feedback. Extensive farming on sloping land can lead to increased soil erosion, which adversely affect land productivity, forcing farmers to reclaim more cropland on even steeper slopes to meet their food needs (Xu et al., 2006). Of course, this will result in even more severe soil erosion.

In contrast, intensive farming can reduce disturbances to sensitive ecosystems. For example, if farming technology improves grain production to an extent such that the existing cropland meets farmers' food needs, cropland expansion will thus be halted; in certain cases, some sloping or inferior cropland may even be converted to forest or grass coverage. Consequently, soil erosion can be mitigated. In this sense, technical change also contributes to environmental improvement, which will benefit grain production in the longer term. Increased grain productivity can further relieve human pressures on cropland expansion, encourage environmental conservation, and finally drive land use onto a sustainable path.[1] That is why it is so crucial to incorporate technical change and environmental consequences into LUCC research.

6.2.2 Empirical Approach

Given that in agriculture, cropland change interacts with farming technology, grain production, and their environmental feedbacks in a complex manner, it becomes plausible and beneficial to represent these interactions and feedbacks with a system of equations. Here we specify four equations for our empirical analysis, with the four dependent variables being cropland use (C) grain production (P), farming technology (T), and soil erosion (S). The interactions of these variables mean that they are endogenous and thus each shows up as explanatory variables in other equations.

The empirical model can be written as:

$$\begin{aligned}
P_{it} &= f_1(X_{it}, T_{it}, S_{it}) + \varepsilon_{it} \\
T_{it} &= f_2(C_{it}, Y_{it}) + \delta_{it} \\
C_{it} &= f_3(T_{it}, P_{it-1}, Y_{it}, Z_{it}) + \upsilon_{it} \\
S_{it} &= f_4(C_{it}, X_{it}, Y_{it}) + \tau_{it}
\end{aligned} \tag{6.1}$$

where f_1–f_4 are functional forms of the four equations, X, Y, and Z are conventional input variables and other socioeconomic and biophysical variables (see detail below), i and t denote the spatial and temporal units of observations, and ε, δ, υ and

---

[1]To be sure, intensive farming has its own environmental problems as well, such as soil salinity and toxicity due to improper chemical applications (Ruttan, 2001).

τ are error terms. Note that cropland area at each time point, instead of cropland change between two points, is used to ensure that the time-series data are consistent for each equation.

Specifically, grain production is defined as a function of farming labor, cropland, other inputs with embedded technical change,[2] and the environmental condition represented by the status of soil erosion and elevation. In our analysis, technical change encompasses three components—fertilizer use, irrigation infrastructure, and multiple cropping. As noted, technical change is affected by resource endowment, changes in relative prices (such as input costs and output prices), and other external socioeconomic and biophysical factors. Therefore, technical change is defined as a function of cropland area, grain price, input cost, farmer net income, market access, and elevation. While cropland area represents resource endowment, elevation denotes the biophysical environment of the observation unit. Farmers' net income affects their abilities to apply technology—a farmer with a higher income has the ability to afford more technical inputs and is more willing to try new farming technologies (Ruttan, 2001). On the other hand, escalating input costs may reduce farmers' incentive for technological adoption because of the application of more technology-embodied inputs (Ruttan, 2001).

Cropland use is determined by grain production, technical change, returns from each land uses, population pressure, and other variables. We use the procurement prices for grain, logs, and livestock to represent the comparative economic returns from cropland, forestland, and grassland. Relative to these market signals, however, decisions on land use in China were and still are influenced by government regulations, such as the food self-sufficiency policy and forest tenure arrangement. The former, reflected in the grain procurement quota, encouraged cropland expansion on slopes previously covered by forest or grassland (Xu et al., 2006). It is hypothesized that a decreasing quota, as a sign of relaxing the policy, should relieve the pressure on cropland expansion. The latter, if clearly defined and enforced, forms the basis for at least stable forest management. In the study area, sloping lands that belonged to the collectives or those without clear ownership, forestland or grassland loss to cropland often occurred (Xu et al., 2006). Soil characteristics affect land suitability for different uses. But measuring soil characteristics for a county is hard because of their variations and data unavailability. So, the average elevation of a county is used as a proxy of soil features as well as other biophysical conditions that affect land use.[3]

Cropland expansion on slopes was deemed as one of the major causes of soil erosion in the Upper Yangtze basin (Xu et al., 2006). Deforestation and forest degradation also damage on ecosystem's abilities to regulate water and soil. On the other

---

[2] Note that technical change and technical adoption are conceptually the same, but they are used interchangeably in this chapter.

[3] Our prior knowledge indicated that elevation, varying from 295 meters (m) to 6109 m for the study region with a mean of 3070 m, is a more meaningful variable, compared to, for instance, slope or range. Because elevation does not change over time, through, it will not be listed in the table of summary statistics of variables below.

hand, the government has been implementing a series of projects to contain soil erosion in the study region. The status of soil erosion is represented by the actual eroded area, which is defined as the function of cropland area, timber harvest, industrial development, implementation of restoration projects, and other biophysical features.

Appropriate estimation method and identification are two critical issues in the following empirical estimation. Since at least one of the explanatory variables in each equation is endogenous and thus correlated with the error term, *OLS* is no longer a valid method to provide unbiased and consistent coefficient estimates. Thus, the 3-stage least squares (*3SLS*) technique is applied in estimating the four-equation system in (6.1). To apply that technique, it is commonly assumed that: (1) all exogenous variables are uncorrelated with any error term at each time point; (2) the covariance matrix of the error terms has a kind of system homoskedasticity; and (3) the rank order condition is satisfied. An equation satisfies the rank order condition if the number of excluded exogenous variables from it is at least as many as the number of the endogenous variables included in it (Wooldridge, 2002). Thus, variables are chosen so as to ensure the identification requirement.

6.2.3 Study Site and Data

The study site was selected along the Jinsha River, part of the upper Yangtze basin with a length of 2,290 km. Included for this study are 31 counties fully located inside the Jinsha River catchment (97.7°–104.8°E, 25.4°–32.7°N), with a total area of 14 million ha. Nineteen counties are in Yunnan province, with an area of 7.2 million ha; and the other 12 in Sichuan, with an area of 6.8 million ha (see Fig. 5.1 of last chapter). The Jinsha River is known for its sharp descent, fragile geological structure, and severe soil erosion. Also, the unfavorable farming condition and poor transportation infrastructure causes the inhabitants of the region to suffer from a high incidence of poverty. Nonetheless, farming has long been the major income source. From 1975 to 2000, the average annual value of farming output accounted for at least 55% of the total agricultural output.[4] Meanwhile, this region plays a critical ecological role in the Yangtze basin because its head waters and primary forests serve important ecosystem functions (Wang & Deng, 2007).

The dataset covers five time points from 1975 to 2000: mid-1970s, mid-1980s, late 1980s, mid-1990s, and late 1990s (hereafter, 1975, 1985, 1990, 1995, and 2000). The land-use data were derived from remote sensing images processed by the Chinese Academy of Sciences (CAS) Liu et al., 2003. The topographic information, such as elevation, also was provided by CAS. The socioeconomic data were from local statistics bureaus or surveys conducted by the authors. Chapter 5 has documented the data details.

In general, cropland is mostly located in the valleys and it has been fairly stable throughout the whole period. However, this aggregation may obscure cropland

---

[4] According to the government statistics, included in agriculture production value are the values of animal husbandry and forestry as well as that of farming.

changes at the county level. Further, opposite land conversions (e.g., from cropland to forest and vice versa) take place simultaneously (see Chapter 5). For example, about 136,000 ha of cropland were converted to grassland, forestland, or other lands from 1975 to 1985; meanwhile, around 147,000 ha of land were converted to cropland from grassland, forestland, and other lands. Notably, the extent of cropland converted to forestland, grassland, and other lands was larger during 1975–1990 than that during 1990–2000. On the other hand, forestland was the largest source of converted cropland, except for the period of 1975–1985 when more grassland became cropland. This situation makes it preferable to identify the LUCC spatial and temporal variations and model their drivers based on county-level data.

Here, grain consists of cereals, beans, and tubers in Chinese statistics. The sown area for grain production accounted for 90% of total sown area of all crops in 1975, and its share was still as high as 77% in 2000. Other inputs include fertilizers, irrigation, and multiple cropping that embody modern farming technologies.[5] In the study region, grain production rose by 70% from 1975 to 2000, with the largest increase taking place between 1990 and 1995 when applications of technical inputs were substantially expanded. In comparison, cropland decreased by 41,000 ha from 1975 to 2000, while farming labor increased by 66%. Yield per farming worker increased by only three percent during that period, leading to a productivity increase of 76% per unit of land. Clearly, production growth cannot be fully explained by land and labor increase; and technology-embodied inputs have made a major difference.

We represent technical change with an index that is a weighted average of the growth rates of the three elements at a given point of time. For instance, the technical change index in 1975 was obtained by multiplying the growth rates of fertilization application, irrigated area, and multiple cropping index in that year. Thus, the larger the derived figure, the greater the technical change rate is. Fertilization application in a county is the amount of fertilizer applied, while the scale of irrigation infrastructure is represented by the irrigation area.[6]

Both grain price and input cost are provincial-level indices. Grain price is the government procurement price index (Chapter 5), while input cost index is an indicator for agricultural input costs given that its calculation takes into account various inputs including fertilizer, machineries, feeds, pesticide, etc. Chomitz and Gray (1996) pointed out that the distance of a parcel of land to roads, representing market access, affects both output and input costs and thus land use patterns. For this study, the total road length in a county is used to reflect transportation cost and market access for the county. The longer the road in a county, the lower transportation costs and the better market access are. As such, the county has more lands allocated to farming. The highway mileage in a county is used to represent road length. Highway rate, the ratio of highway length to county area, is the variable used in the model to remove the effect of county size.

---

[5] Multiple cropping refers to the situation where the cultivated land is used more than once a year. It is thus measured with the ratio of the total sown area divided by the total cultivated area. As such, it enhances the land-use intensity.

[6] Field visits indicate that plastic sheeting in high elevations can effectively raise soil temperature and maintain moisture, resulting in greater probability of crop success and higher grain yield.

The eroded area in a county was estimated by local water resource bureaus based on their field surveys. The Erosion Control Project along the Upper Yangtze, carried out since the late 1980s, is the largest of such projects along the Jinsha River (Wang, 2003), and it combines biological, engineering, and tilling measures. Also, the Upper Yangtze forestation project launched in 1989 listed erosion control as one of its objectives. Table 6.1 summarizes all the variables.

Table 6.1 Summary Statistics of Variables Used in the Structural Model

	Measurement	1975	1985	1990	1995	2000
Endogenous						
Cropland area	Ha	1,603,687	1,614,621	1,561,700	1,563,426	1,560,719
Grain production	Ton	1,623,484	1,953,656	1,941,526	2,632,755	2,786,049
Technology index		0.073	0.044	0.054	0.065	0.035
Soil erosion area	1,000 ha	2,033	2,385	2,894	3,011	2,743
Exogenous						
Population density	Person/km^2	0.66	0.73	0.78	0.82	0.86
Agricultural labor	1,000 Person	2,178	2,798	3,047	3,509	3,623
Irrigation area	Ha	189,588	192,676	194,511	212,489	269,180
Fertilizer	Ton	35,984	62,444	72,991	140,065	215,603
Multiple cropping index		1.397	1.390	1.448	1.623	1.707
Grain price[1]	Index (1990 = 1)	0.587	0.669	0.972	1.619	2.035
Log price[1]	Index (1990 = 1)	0.589	0.531	0.927	1.046	1.294
Livestock price[1]	Index (1990 = 1)	0.860	0.863	1.085	1.873	2.218
Industrial output (IO)[2]	1,000 RMB	69,534	141,243	169,494	516,442	698,396
Agricultural output (AO)[2]	1,000 RMB	189,063	329,188	351,141	533,255	729,546
IO/AO		0.320	0.403	0.443	0.768	0.762
Grain procurement quota[3]	Ton	170,423	141,292	111,231	115,691	107,481
Highway rate	km/ha	0.0009	0.0014	0.0016	0.0021	0.0031
Farmers' income	RMB	81	278	344	565	1,132
Cost index[1]	Index (1975 = 100)	100.100	107.127	136.255	238.744	361.973
Timber harvests	m^3	259,925	833,007	850,862	854,845	380,868
Erosion control project	Dummy	=1, if project implemented; = 0, otherwise				
Forestation project	Dummy	=1, if project implemented; = 0, otherwise				
NFPP project	Dummy	=1, for the year of 2000; = 0 for other years				

[1] Grain, log, and livestock price indices are provincial aggregates.
[2] Agricultural and industry outputs are output values at 1990 constant price.
[3] Grain quota is from the local grain bureau, and soil erosion area is from the local Water Resources Bureau. All other data come from local statistics bureaus and government documents.

6.3 Estimated Results

Table 6.2 lists the estimated results. Yearly and provincial dummy variables are included in each equation to control the temporal trend and regional heterogeneity, but results are omitted to save space. The coefficient for each variable in the grain production, cropland, and soil erosion equations is the corresponding elasticity, not the partial effect. This presentation can indicate the extent of driving forces' effects on each of the dependent variables. For the farming technology equation, a coefficient value is the estimated partial effect because the technical change itself is a variable of percentage. In general, the Chi-square test shows that every equation is significant, meaning that there is a significant relationship between all independent variables with dependent variable. R^2 values further indicate that the independent variables in grain production and cropland equations fit very well. Many variables are statistically significant with expected signs. When those four dependent variables are used as independent variables in other functions, they are all statistically significant. This proves our claim that interactions and feedbacks exist among cropland use, grain production, farming technology, and environmental consequence.

Results for grain production are mostly as expected. Land, labor, and technical input all have significant positive effects, while deteriorated soil condition has a significant negative effect on production. Grain production increases by 0.6% and 0.27% with one percent increase of cropland area and labor, respectively, holding other variables constant. One percent increase of fertilizer application leads to 0.16% increase of grain production, and one percent increase of the multiple cropping index results in a 0.26% increase in grain production. The effect of irrigation is not significant, probably because its effect is partially captured by the fertilizer variable. Irrigation promotes the application of fertilizer because they work more effectively with adequate moisture (Wang & Deng, 2007). Soil erosion negatively affects production. Holding other variables constant, if a county has 1,000 more ha of eroded land, grain production is reduced by 0.1%. The effect of elevation is positive on grain production. Although different from our expectation, its magnitude is small.

Most variables in cropland equation also have the expected effects. Livestock price, highway density, and grain procurement quota have significant positive effects on cropland expansion, whereas cropland area decreases significantly with altitude and technical change. That is, cropland area increases by 0.55% when the highway in a county increases by one percent, indicating that better road access leads to more crop production and thus more demand for cropland. One percent increase of per capita grain procurement quota leads to 0.06% increase in cropland, and cropland area in a county decreases with rising elevation. The effect of livestock price increase on cropland expansion is relatively large: one percent of livestock price increase causes cropland to expand by 2.6%. This suggests that some crops are planted for animal feeding stocks, such that a higher livestock price drives more cropland use to expand livestock production. Animal husbandry is indeed regarded as a major means of improving income and alleviating poverty in many mountainous regions (Wang & Deng, 2007). One of the most interesting results for the cropland

Table 6.2 Estimated Results of the Structural Model for Cropland Use

	Grain production	Cropland use	Farming technology	Eroded area
Grain production		0.392 (0.002)***		
Cropland use	0.604 (0.062)***		−0.198 (0.064)***	0.380 (135.085)***
Farming technology		−0.035 (0.353)**		
Eroded area	−0.099 (0.041)**			
Agri. labor	0.267 (0.052)***			
Fertilizer	0.158 (0.027)***			
Irrigation	0.039 (0.034)			
Multiple cropping	0.255 (0.070)***			
Log price		−0.073 (0.998)		
Livestock price		2.588 (1.067)*		
Grain price		−1.322 (0.605)	0.138 (0.078)*	
Population density		0.076 (0.699)		
Grain procurement quota		0.058 (1.847)*		
Highway rate		0.547 (0.156)***	0.095 (0.047)**	
Farmers' income			0.008 (0.004)*	
Cost index			−0.119 (0.049)**	
Forest ownership		0.223 (0.196)***		
Industrial development		0.013 (0.063)		0.049 (133.162)
Erosion control project				−0.165 (228.745)**
Forestation project				−0.114 (230.160)
Timber harvests				0.156 (0.003)***
Elevation	0.101 (0.028)***	−1.000 (0.055)***	0.002 (0.003)	−0.114 (0.097)
R^2	0.98	0.80	0.17	0.49

Note:
1. Numbers in parentheses are standard error of the coefficient.
2. There were 98 observations used, because of some missing values for variables.
3. The level variables were log transformed before estimation.
"*", "**" and "***" represent 10, 5 and 1% significance level, respectively.

equation is that technical change has a significant effect on reducing cropland expansion. Controlling other factors, cropland area decreases by 0.04% with one percent increase in the technical change index.

The effects of the explanatory variables on technical change conform to our expectations to a large extent as well. Farmers' income, grain price, and road length all have significant positive effects on technical change, while higher input costs and abundant cropland resource reduce their incentives to apply technologies intensively. Despite its significance, the magnitude of the income coefficient is small. Holding other factors constant, the rate of technical change is expected to increase by 0.007 if farmers' income increases by 100 yuan. One unit increase of grain price induces a 0.14% increase of technical change, and one thousand kilometers of newly constructed highway in a county result in a 0.09% increase in technical change. On the other hand, one unit increase in agricultural input cost causes the rate of technical change to drop by 0.12%. The adoption rate of land-intensive or land-saving technology is 0.198% lower if the cropland area in a county increases by 10,000 ha. This result further validates our hypothesized interaction between technical change and cropland use. Although most variables have significant effects on technical change, the overall explanatory power of this equation is poor—only 17% of the variation of technical change.

Eroded area is significantly related to human activities, and the implementation of conservation projects effectively reduces it (Yin et al., 2005). On average, erosion control projects reduce around 470 ha of eroded area in a county, controlling the effects of other variables. Afforestation and reforestation also have the same effect on erosion control, but it is insignificant, probably in part because forestation projects were introduced much later than the engineering ones (Yin et al., 2005). In addition, cropland expansion has significant effect on erosion deterioration. One percent of cropland increase induces 386 more ha of eroded land. Meanwhile, timber harvests also intensify soil erosion. One percent increase in timber harvest causes eroded areas to increase by 0.16%, holding other variables constant.

6.4 Conclusions and Implications

This chapter has developed a structural model to capture the interactions and feedbacks in the dynamic process of cropland change, based on the experience in the upper Yangtze basin of China. We proposed and proved that cropland use interacts with grain production and agricultural technology and that the environmental consequence of the land use also imposes significant feedbacks. Our results demonstrate the critical role of technical change in providing food on limited cropland. Counties with greater technical change have less land allocated to grain production, leading to more intensified use of existing cropland and the conversion of marginal cropland to other uses. Thus, it can be inferred that when the increased productivity resulting from technical change can satisfy the food supply from a limited land base, farmers will transfer labor and land resources to other activities to earn more income. Controlling cropland expansion reduces the extent of soil erosion, which then benefits grain production. The study also highlights the importance of institutions and policies in environmental protection and sustainable land use. For instance, the food self-sufficiency policy causes cropland expansion, especially in counties where cropland resources are very scarce. While timber harvests induce more soil erosion, erosion control projects can effectively reduce the eroded area.

These and other results not only improve our understanding of the complex human and natural connections in LUCC, but also carry significant policy implications for sustainable land use. First, technical change and technology innovation should be encouraged and supported by governments at the national and local levels. Higher input costs and lower crop prices compared to off-farm wages discourage the adoption of technology by individual farmers. And our data show that the rate of technical change was slowing down from 1975 to 2000—the conventional technology is reaching its maximum potential, and the amount of fertilizers or irrigation cannot be applied limitlessly. Technical innovation and adoption have thus become even more important to continued grain production. The policy environment should be improved to promote extension services, enhance the distribution network of inputs, and strengthen investment in agricultural research and development.

Further, as more and more efforts are devoted to increasing farmers' income from livestock-based activities, the challenge to balance cropland expansion for feed stock production and natural resource protection becomes more acute. This is because a higher livestock price induces more animal production, which leads to greater demand for cropland to provide more feed stock (e.g., corn and beans). Expanded cropland will cause more soil erosion and other environmental problems. Our work also justifies the environmental protection and ecological restoration projects undertaken by the Chinese government. Meanwhile, the role of the market should be respected when it can exert its function in efficient resource allocation. For example, when farmers are subsidized to plant trees, it is not necessarily the right move for the government to dictate the decisions of tree planting and harvesting (Yin et al., 2005). Even if certain species and management practices may not be the best from the perspective of environmental protection, government dictation can lead to worse outcomes. Rather, the government ought to provide farmers with the needed market information and technical know-how. Also, the government should let farmers face the market prices and possess secure use and disposal rights for their trees. Farmers can make rational long-run decisions about land use based on the market signals and thus achieve more sustainable land use.

Timber harvest has a significant negative effect on soil erosion. The "logging bans" policy effectively reduced the commercial timber production in the late 1990s. However, timber production amounts to only a small percentage of the total forest removals. Fuelwood consumed by rural households and logs used for local construction take up a large share of the annual resource consumption (Xu et al., 2006). Therefore, in addition to reducing commercial timber production, the use of electricity, biogas, and fuel-efficient stoves in the rural area can reduce the demand for fuel wood. The government should invest more in such technologies and encourage their adoption.

It should be noted that while this study has generated some important results, more needs to be done in future research. First, LUCC are regarded not only as a driver of climate change, climate change can in turn impact LUCC (USGCRP, 2003) through hydrological and terrestrial biological systems. Thus, incorporating the effects of climate change and other biophysical factors into the land-use change model will allow us to reflect natural and human interactions more comprehensively. Second, in view of real-world complications, this study has focused only on the multiple facets of agriculture and cropland use in the structural model. Such a systematic method can be applied to other land uses. Moreover, the connections among these models should be established. In doing so, we will gain a more complete picture of the dynamics of all major land categories. Finally, the characteristics of LUCC may be different at different scales (e.g., county vs. household scale). The broad scale LUCC analysis will obscure the variation at the finer scale; on the other hand, the land use changes at the fine scale will result in environmental and economic changes that can only be fully appreciated at the broad scale. Also, the effects of policy or institutional factors may vary at the different scales. Thus, it is worthwhile to conduct LUCC research at multiple scales and to examine the LUCC driving forces in a spatially explicit way.

Acknowledgments This article has been accepted for publication in *Sustainability Science*. The authors appreciate comments made by the journal reviewers and Editor, as well as Jiaguo Qi, Larry Leefers, Karen Potter-Witter, and other colleagues. They also are grateful to the US National Science Foundation for funding.

References

Chomitz, K. M., & Gray, D. A. (1996). Roads, land use, and deforestation: A spatial model applied to Belize. *World Bank Economic Review, 10*(3), 487–512.

Foley, J. A., DeFries, R., Asner, G. P., Barford, C., Bonan, G., Carpenter, S. R., et al. (2005). Global consequences of land use. *Science, 309*, 570–574.

Geoghegan, J., Villar, S. C., Klepeis, P., Mendoza, P. M., Himmelberger, Y. O., Chowdhury, R. R., et al. (2001). Modeling tropical deforestation in the southern Yucatán Peninsular region: comparing survey and satellite data. *Agriculture, Ecosystems and Environment, 85*, 25–46.

Irwin, G. E., & Geoghegan, J. (2001). Theory, data, methods: Developing spatially explicitly economic models of land use change. *Agriculture Ecosystems and Environment, 85*(1), 7–23.

Kaimowitz, D., & Angelsen, A. (1998). *Economic models of tropical deforestation: A review.* Bogor, Indonesia: Center for International Forestry Research.

Lambin, E. F., Geist, J. H., & Lepers, E. (2003). Dynamics of land-use and land-cover change in tropical regions. *Annual Review of Environment and Resources, 28*, 205–241.

Lambin, E. F., Rounsevell, M., & Geist, H. (2000). Are agricultural land-use models able to predict changes in land use intensity? *Agriculture, Ecosystems & Environment, 82*(1–3), 321–331.

Lin, J. Y. (1992). Rural reforms and agricultural growth in China. *American Economic Review, 82*(1), 34–51.

Liu, J. Y., M. L. Liu, D. F. Zhuang, Z. X. Zhang, and X. Z. Deng. (2003). Study on spatial pattern of land-use change in China during 1995–2000. Science in China (Series D) *46*(4): 373–384.

Millennium Ecosystem Assessment. (2005) *MA Major Findings*. www.millenniumassessment.org

Müller, D., & Zeller, M. (2002). Land use dynamics in the central highlands of Vietnam: A spatial model combining village survey data with satellite imagery interpretation. *Agricultural Economics, 27*, 333–354.

Ruttan, V. W. (2001). *Technology, growth, and development: An induced innovation perspective.* Oxford, UK: Oxford University Press.

Turner II, B. L., Lambin, E. F., & Reenberg, A. (2007). The emergence of land change science for global environmental change and sustainability. *PNAS, 104*(52), 20666–20671.

USGCRP. (2003). *Land use/land cover change: USGCRP program element.* Retrieved from http://www.usgcrp.gov/usgcrp/ProgramElements/land.htm

Veldkamp, A., & Fresco, L. O. (1996). CLUE: a Conceptual model to study the conversion of land use and its effects. *Ecological Modeling, 85,* 253–270.

Walker, R., & Moran, E. (2000). Deforestation and cattle ranching in the Brazilian Amazon: External capital and household processes, *World Development, 28*(4), 683–699.

Wang, X. T. (2003). *Building an ecological shield along the Upper Yangtze river: Priorities and measures.* Beijing, China: China Agriculture Publishing House.

Wang, Y. K., & Deng, Y. L. (2007). *An assessment of sediment retention capacity of ecosystem in the upper Yangtze basin* (report). Chengdu, Sichuan: Institute of Mountain Hazards and the Environment, Chinese Academy of Science.

Wooldridge, J. M. (2002). *Econometric analysis of cross section and panel data.* Cambridge, MA: The MIT Press.

Xu, J., Yin, R. S., Li, Z., & Liu, C. (2006). China's ecological rehabilitation: Progress and challenges. *Ecological Economics, 57*(4), 595–607.

Yin, R. S., & Hyde, W. F. (2000). Trees as an agriculture sustaining activity: The case of northern China. *Agroforestry Systems, 50,* 179–194.

Yin, R. S., Xu, J. T., Li, Z., & Liu, C. (2005). China's ecological rehabilitation: Unprecedented efforts, dramatic impacts, and requisite policies. *China Environment Series, 6,* 17–32.

Chapter 7
Quantifying Terrestrial Ecosystem Carbon Dynamics in the Upper Yangtze Basin from 1975 to 2000

Shuqing Zhao, Shuguang Liu, Runsheng Yin, Zhengpeng Li, Yulin Deng, Kun Tan, Xiangzheng Deng, David Rothstein, and Jiaguo Qi

Abstract Quantifying the spatial and temporal dynamics of carbon stocks in terrestrial ecosystems and carbon fluxes between the terrestrial biosphere and the atmosphere is critical to our understanding of regional patterns of carbon storage and loss. Here we use the General Ensemble Biogeochemical Modeling System to simulate the terrestrial ecosystem carbon dynamics in the Jinsha watershed of China's upper Yangtze basin from 1975 to 2000, based on unique combinations of spatial and temporal dynamics of major driving forces, such as climate, soil properties, nitrogen deposition, and land use and land cover changes. Our analysis demonstrates that the Jinsha watershed ecosystems acted as a carbon sink during the period of 1975–2000, with an average rate of 0.36 Mg/ha/yr, primarily resulting from regional climate variation and local land use and land cover change. Vegetation biomass accumulation accounted for 90.6% of the sink, while soil organic carbon loss before 1992 led to lower net gain of carbon in the watershed, and after that soils became a small sink. Ecosystem carbon sink/source pattern showed a high degree of spatial heterogeneity. Carbon sinks were associated with forest areas without disturbances, whereas carbon sources were primarily caused by stand-replacing disturbances. This highlights the importance of land-use history in determining the regional carbon sink/source pattern.

Keywords General Ensemble Biogeochemical Modeling System (GEMS) · Carbon flux · Carbon stock · Climate change · Land use and Land cover change (LUCC) · Jinsha watershed

S. Zhao (✉)
Department of Forestry, Michigan State University, East Lansing, MI 48824, USA;
Center for Global Change and Earth Observations, Michigan State University,
East Lansing, MI 48823, USA
e-mail: szhao@usgs.gov
Ecosystem Policy Institute of China, Chaoyang District, Beijing 100102, P.R. China
e-mail: yinr@msu.edu

R. Yin (ed.), *An Integrated Assessment of China's Ecological Restoration Programs*,
DOI 10.1007/978-90-481-2655-2_7, © Springer Science+Business Media B.V. 2009

7.1 Introduction

The terrestrial carbon budget results from complex interactions and feedbacks among plant productivity, decomposition, climate, soil properties, and human activities. These biogeochemical and biophysical processes occur across boundaries of traditional disciplinary investigations and on multiple time scales (Bala et al., 2007; Chapin et al., 2006). Therefore, to comprehensively understand the causes and magnitudes of ecosystem carbon fluxes, and hence ecosystem's carbon storage, it is critical to study the systems in meaningfully large units and over sufficiently large-time scales.

However, there have been two major challenges in estimating ecosystem carbon dynamics over large spatiotemporal scales. First, quantifying the carbon exchanges between the terrestrial biosphere and the atmosphere due to land use change is still the largest uncertainty in the global and regional carbon budget (Canadell, 2002; Achard, Eva, Mayaux, Stibig, & Belward, 2004; Ramankutty et al., 2007). Land use and land cover change (LUCC), including land conversion from one type to another and land cover modification through land-use management, altered a large proportion of the earth's land surface (Meyer & Turner, 1992; Vitousek, Mooney, Lubchenco, & Melillo, 1997; Foley et al., 2005; Zhao et al., 2006), and disturbed the biogeochemical interactions between the terrestrial biosphere and the atmosphere (Schimel et al., 2001; Houghton & Goodale, 2004). From 1850 to 2000, roughly 35% of global anthropogenic CO_2 emissions resulted directly from land use changes (Houghton, 2003), whereas contemporary land use changes are considered to be the dominant driver for some regional terrestrial carbon sinks, contributing to a large portion of current northern hemisphere terrestrial sinks (Fang, Chen, Peng, Zhao, & Ci, 2001; Fang, Oikawa, Kato, Mo, & Wang, 2005; Choi, Lee, & Chang, 2002; Kauppi et al., 2006). This highlights the importance of incorporating spatially explicit LUCC information into the estimation of regional carbon budgets.

Secondly, it has been a primary challenge to scale up the carbon fluxes and stocks measured or simulated at the site level to a regional level, mainly due to the heterogeneity of a range of environmental variables driving ecosystem processes (Jenkins, Birdsey, & Pan, 2001; Tickle, Coops, & Hafner, 2001; Binford, Gholz, Starr, & Martin, 2006; Gimona, Birnie, & Sibbald, 2006). Integrating ecosystem models with geographic information systems (GIS) has provided the capability for extrapolating local information to a wider region. However, many previous models for estimating regional carbon fluxes are based on the direct application of site-scale methods over grid cells larger than site plots (e.g., Potter et al., 1993; Pan et al., 1998; Tian et al., 1998), and are thus unable to capture the influences of heterogeneous environmental conditions at finer spatial scales. This can lead to significant biases both in scientific research and policy decision making.

In this chapter, we quantify the terrestrial ecosystem carbon dynamics in the Jinsha watershed of China's upper Yangtze basin from 1975 to 2000, using the General Ensemble Biogeochemical Modeling System, which is capable of dynamically assimilating LUCC information into the simulation process across large spatial extents, and of upscaling carbon stocks and fluxes from local to regional levels (Liu, Kaire, Wood, Diallo, & Tieszen, 2004; Liu, Loveland, & Kurtz, 2004). GEMS is driven by the spatial and temporal Joint Frequency Distribution of major driving variables, such as climate, soil properties, nitrogen deposition, land cover changes, and land use practices with a range of resolutions.

7.2 Methods

7.2.1 Site Description

The Jinsha watershed is situated at the upper reaches of the Yangtze River (Fig. 7.1). It comprises 37 counties across Sichuan Province, Yunnan Province and the Tibet Autonomous Region, with an area of 171,000 km^2 (97° 40'–104°48'E, 25°18'–32°48'N). Elevations within the watershed range between 500 and 6740 m. Its subtropical climate has an average annual rainfall of 800–1,000 mm and temperatures of 14–18C$^\circ$ (Lu, 2005). The region features a wide variety of ecosystems that have been recognized as a great biodiversity hotspot (Conservation International, 2002). Accompanying the diverse ecosystems are the extreme fluctuations in topography and landscapes, which make the regional environment vulnerable. But it is the human land-use practices and the climate variation

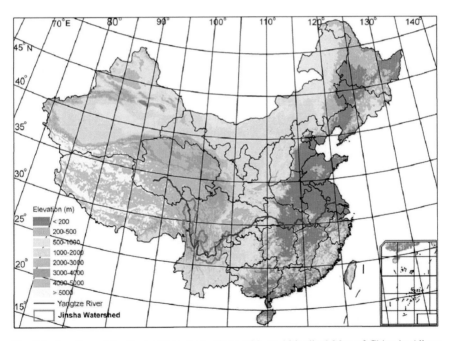

Fig. 7.1 Location of the Jinsha Watershed, Along with an Altitudinal Map of China in Albers Equal-Area Conic Projection

that have altered the regional environment (Xiang, Yin, Xu, & Deng, 2008). There-fore, it is important to understand how the regional LUCC and climate variation have affected terrestrial ecosystem carbon dynamics.

7.2.2 Model Description

General Ensemble Biogeochemical Modeling System (GEMS) is a modeling system that was developed to better integrate well-established ecosystem biogeochemical models with various spatial databases for simulations of biogeochemical cycles over large areas. GEMS consists of three major components: one (or multiple) encapsu-lated ecosystem biogeochemical model, an automated model parameterization sys-tem (AMPS), and an input/output processor (IOP) (see Fig. 3.2). The plot-scale Erosion-Deposition-Carbon Model serves as the encapsulated ecosystem biogeo-chemical model in GEMS (Liu, Bliss, Sundquist, & Huntington, 2003). The spatial deployment of the site-scale model in GEMS is based on the spatial and tempo-ral joint frequency distribution of major driving variables (e.g., land cover and land use change, climate, soils, disturbances, and management). Land cover sequences in GEMS are generated by filling the land cover gaps between consecutive land cover maps using a Monte Carlo approach. For details on the model description, see Liu (2008).

7.2.3 Environmental Variables

The LUCC data The LUCC data in the Jinsha watershed between 1975 and 2000 were obtained using cloud-free Landsat Multispectral Scanner (MSS) and Thematic Mapper (TM) remote sensing images. The data covered 5 periods: 1975, 1985, 1990, 1995, and 2000. Land-cover data with a resolution of 100×100 m were divided into six types: cropland, grassland, shrub, forest, disturbed or transitional (i.e., land in an altered unvegetated state which is in transition from one cover type to another), and others (e.g., urbanized area, body of water, snow, ice, and barren land).

Climate data Long-term monthly minimum temperature, monthly maximum temperature, and mean monthly precipitation were obtained from a 1961 to 1999 climate database of China at $0.1° \times 0.1°$ resolution, generated from 680 climatic sta-tions across the country (Fang, Piao, Tang, Peng, & Wei, 2001).

Soil properties Soil data were taken from a national soil database with multiple layer information at 10×10 km resolution developed by Shi and Yu (2002). The soil properties that we use include soil texture (sand, silt, and clay fraction), bulk density, organic matter content, wilting point, and field capacity by each layer. Soil drainage classes, from excessively well drained to very poorly drained, came from a GIS-derived integrated moisture index (Iverson, Dale, Scott, & Prasad, 1997), which were generated by integrating hill shade, flow accumulation, curvature, and water-holding capacity based on digital elevation model (DEM, with a resolution of 100×100 m) and soil texture data.

Nitrogen deposition The information was downloaded from http://eos-webster. sr.unh.edu/data_guides/china_dg.jsp, which was created by Changsheng Li and Steve Frolking of the Complex Systems Research Center (Institute for the Study of Earth, Oceans, and Space, Morse Hall, University of New Hampshire, Durham, New Hampshire, USA). It covers both wet and dry sources at the county level.

Forest inventory Species composition and forest age and biomass distribution data at the county level were obtained from the National Forest Resource Inventory database for Sichuan, Yunnan, and Tibet. Initial data used for each county fall within the 1970s and 1980s, which generally correspond to the start time of this study. Timber volume was converted to biomass for each forest species based on the biomass expansion factor method developed by Fang, Chen, et al. (2001).

7.3 Results and Discussion

7.3.1 Land Cover Changes

There was no single dominant land cover in the study area (Table 7.1). Forest (37%), grassland (32%), shrub land (13%), and cropland (11%) were the four major land cover types, covering about 94% of the region. From 1975 to 2000, little significant change was observed on the composition of land cover types (Table 7.1). Disturbed landscape, primarily caused by forest harvesting, decreased from 0.6% in 1975 to 0.2% in 2000, suggesting a reduction in forest clear-cutting activities during the period.

However, the region-wide summary of land cover composition (Table 7.1) does not reflect the changes that happened at finer spatial scales. It is very clear that although there were not significant net changes in land covers at the regional scale across the study period, various land covers did change at finer scales (Fig. 7.2). This highlights the necessity of characterizing land cover changes using spatially explicit information at the scale where land cover change activities happen (i.e., at the field scale). The land cover transitions reveal detailed and rather dynamic land cover changes between two neighboring periods (Fig. 7.2). The total areas that changed between two consecutive periods varied between 11.6 to 10.2%, with a slight decreasing trend. The cumulative land conversion line for the period of 1985–1990

Table 7.1 Land Cover Proportions (%) in the Jinsha Watershed During the Period of 1975–2000

Land cover category	1975	1985	1990	1995	2000
Cropland	11.1	11.1	10.9	10.9	10.9
Grassland	32.2	32.4	32.6	32.7	32.8
Shrub	13.5	13.3	13.2	13.3	13.4
Forest	37	37.4	37.6	37.7	37.8
Disturbed	0.6	0.3	0.2	0.2	0.2
Other land cover	5.7	5.5	5.5	5.3	5.0

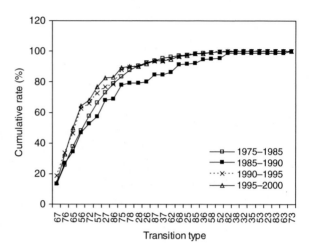

Fig. 7.2 Land Cover Transitions Between Two Consecutive Periods from 1975 to 2000. Cumulative Rates Are Expressed as the Cumulative Percentage of Total Changed Area for Each Consecutive Period. Land Cover Code of 2, 3, 5, 6, 7, and 8 Denotes Other Land Cover, Disturbed Land, Shrub, Forest, Grassland, and Cropland, Respectively

is below those for the other three periods, suggesting that land cover transitions between 1985 and 1990 were more diverse and significant.

The major changes, which accounted for more than 70% of the total, were between forest and grassland, forest and shrub, grassland and miscellaneous land covers, grassland and shrub, and cropland and forest for all four consecutive periods (Fig. 7.2). Even though the transitions from forest to grassland or from grassland to forest were one of the major land cover changes, the net change from 1975 to 2000 was small. Net transition from forest to shrub was 1% of the total area from 1975 to 2000, indicating a net loss of forest to shrub due to logging. At the same time, net transition from cropland to forest was 0.6% of the total area, resulting from afforestation efforts in the region.

7.3.2 Comparison of Simulated and Observed Crop Yield

Figure 7.3 shows the comparison of temporal changes of GEMS simulated crop yields and yields obtained from statistical yearbooks for one county in the Jinsha watershed (Panzhihua) from 1986 to 2000 (ECPSY, 2002). It can be seen that model simulations were in general agreement with statistical yearbook data.

7.3.3 Changes in Ecosystem Net Primary Production (NPP)

From 1975 to 2000, ecosystem NPP in the Jinsha watershed increased from 281.8 to 372.5 g C/m^2/yr, with a growth rate of 32.2% (Fig. 7.4a). Inter-annual variability in NPP corresponded closely with the variations in climate. Over the period of 1975 to 2000, both annual precipitation and mean annual temperature significantly increased (Fig. 7.4b). NPP was anomalously low in 1978, 1992, and 1996, and high

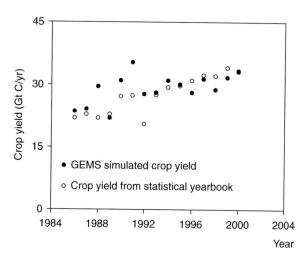

Fig. 7.3 A comparison of GEMS-Simulated Crop Yield and Crop Yield Obtained from the Statistical Yearbook for One County in the Jinsha Watershed (i.e., Panzhihua) from 1986 to 2000. Total Crop Yield was Converted to Dry Weight by Multiplying a Conversion Factor of 0.85, and then to Carbon by Further Multiplying a Factor of 0.45. Included Crops Are Wheat, Corn, Soybean, Tuber, Fiber, and Rapeseed (ECPSY, 2002)

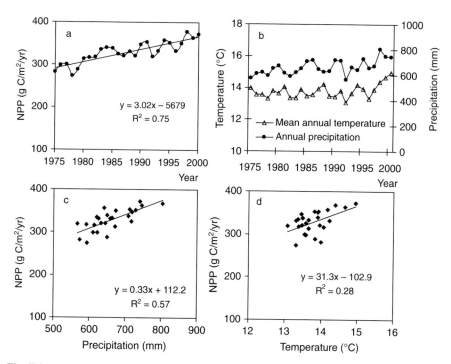

Fig. 7.4 The Relationship Between Net Primary Productivity (NPP) and Climate in the Jinsha Watershed Between 1975 and 2000. (**a**) The Temporal Change of NPP; (**b**) The Temporal Changes of Annual Precipitation and Mean Annual Temperature; (**c**) The Relationship Between NPP and Annual Precipitation; and (**d**) The Relationship Between NPP and Mean Annual Temperature. $P <$ 0.05 in Each Case

in 1998 (Fig. 7.4a), years that are strongly associated with changes in climate. Low NPP in 1978, 1992, and 1996 was coupled with low precipitation and cooling temperature, whereas the high NPP in 1998 was associated with anomalously high precipitation (Fig. 7.4b). The relationships between NPP and climate variables indicate that ecosystem NPP is closely correlated with annual precipitation and mean annual temperature, and the association with precipitation is stronger than that with temperature (Fig. 7.4c, d). Climate-driven increases in ecosystem production have been reported in many previous studies (e.g., Fang et al., 2003; Nemani et al., 2003; Cao, Prince, Small, & Goetz, 2004).

7.3.4 Changes in Carbon Stocks

Ecosystem carbon stock density, including biomass carbon and soil organic carbon (SOC), increased from 112.7 Mg/ha in 1975 to 122.2 Mg/ha in 2000, with a growth rate of 8%. Both biomass carbon (including live and dead vegetation biomass) and SOC showed a similar enhanced trend but a difference in magnitude, with a growth rate of 44 and 1%, respectively (Fig. 7.5a). It is well known that

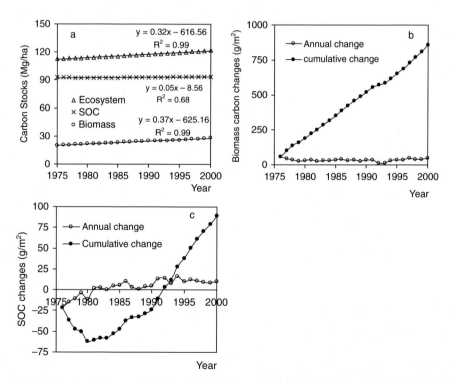

Fig. 7.5 Temporal Changes of Carbon Stocks Between 1975 and 2000 in the Jinsha Watershed. (**a**) The Overall Ecosystem Dynamics ($P < 0.05$); (**b**) Annual and Cumulative Changes in Biomass Carbon; and (**c**) SOC

increases in plant productivity will lead to a greater storage of carbon in vegetation biomass, especially in ecosystems dominated by woody vegetation (Rustad et al., 2001). Figures 7.5b and 7.6c show the relative contributions of biomass carbon and SOC to the accumulation of ecosystem carbon stock. The annual change of biomass carbon (i.e., biomass carbon of current year minus that of previous year) varied from 13.2 to 55.5 g/m^2/yr. The cumulative temporal changes of biomass carbon were all positive, and increased over time, indicating that vegetation biomass consistently gained C during the period of 1975 and 2000, and that the magnitude grew over time. The cumulative increase of biomass carbon amounted to 857.3 g/m^2 by 2000 (Fig. 7.5b). In contrast, the annual change of SOC was negative before 1981, and it became positive thereafter. Consequently, the cumulative change of SOC decreased from −21.4 g/m^2 in 1976 to −62.1 g/m^2 in 1980, then increased from −60.6 g/m^2 in 1981 to 89.2 g/m^2 in 2000, with the inflection time point of 1992 (shifting to positive) (Fig. 7.5c). This variation suggests that soils in the Jinsha watershed ecosystem changed from a net carbon source to a net carbon sink during the period of 1975 and 2000.

The results also indicate that the increase in ecosystem carbon stock from 1975 to 2000 was primarily augmented by biomass carbon accumulation, and SOC loss before 1992 led to a lower net gain of carbon in the watershed. The storage of carbon in soils results from the balance between annual plant detritus input to the soil and the decomposition of soil organic matter. It is commonly observed that the decomposition process is more sensitive to environmental change than plant production and its consequent debris input to the soil (Jenkinson, Adams, & Wild, 1991; Davidson & Janssens, 2006). Thus, the response of soil to environmental change (e.g., climate or disturbance) often results in a net release of carbon to the atmosphere by the terrestrial ecosystem or at least less net carbon sequestration from the atmosphere by the terrestrial ecosystem.

7.3.5 Carbon Sink/Source Pattern Between 1975 and 2000

Although ecosystem NPP and carbon storage showed an overall increase from 1975 to 2000 in the Jinsha watershed, the carbon sink/source pattern had a high degree of spatial heterogeneity. Carbon sink/source strength over the period of 1975 and 2000 was derived from the difference between the 3-year mean ecosystem carbon stock for 1975–1977 and that for 1998–2000, divided by 25, which was equal to the net biome productivity (NBP) in Carbon-Cycle Concepts and Terminology (Chapin et al., 2006).

Carbon sources were mostly concentrated in the northwestern portion, especially within the two counties in Tibet (i.e., Jiangda and Mangkang), although they were scattered across the study area. The south-central portion was primarily a carbon sink, and the east was close to neutral except for some local "hot-spots" of carbon sinks. Overall, 20.6% of the total area acted as a carbon source during the period, 33.1% was carbon neutral, and carbon sequestration occurred in 46.3% of the study area. The magnitude of carbon sink mainly fell in the range of 0–2 Mg/ha/yr (Fig. 7.6a).

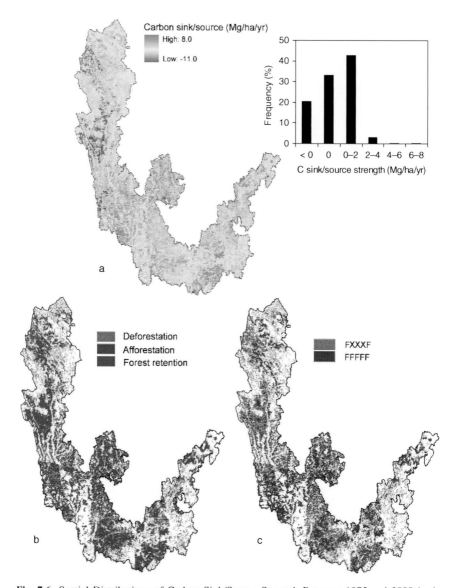

Fig. 7.6 Spatial Distributions of Carbon Sink/Source Strength Between 1975 and 2000 in the Jinsha Watershed. (**a**) The sink/source strength over Time, with the *Inset* Graph Denoting the Area Frequency Distribution of Carbon Sink/Source Strength; (**b**) Afforestation and Deforestation Between 1975 and 2000; and (**c**) Consecutive Forest Cover Changes for Five Time Periods, with F and N Representing Forest and Non-Forest. X in XXX Could Be Either F or N, but at Least with One N

As discussed earlier, the increase in ecosystem carbon stock over time was primarily driven by the accumulation of vegetation biomass carbon. Forest was the absolute dominant component of vegetation biomass. Therefore, we overlaid the land cover maps of 1975 and 2000 to investigate the possible relationship between forest change and ecosystem carbon sink/source pattern. Our results show that carbon sinks were generally associated with forest areas, and that local "hot-spots" of carbon sinks and sources tend to correspond to afforested and deforested areas, respectively. The non-forested areas were generally carbon neutral (Fig. 7.6b). However, carbon sources concentrated in the northwestern portion were also located in the forest retention area. These seemingly conflicting trends may be explained by further exploring what happened to the retained forests within the period. So, we investigated the consecutive cover changes of the retained forests for the five time periods (Fig. 7.6c). It is found that carbon sinks were associated with forest areas with little disturbance (that is, consistently being a forest area throughout the five periods, FFFFF), whereas C sources primarily resulted from clear-cutting and/or shifting cultivation (that is, from forest to non-forest and then back to forest, or FXXXF).

Although the northwestern portion of the study area was covered by forest both in 1975 and 2000, dramatic disturbances occurred to the forested land within the period. Stand-replacing forest harvesting altered forest age; thus the age of the forest in 2000 was a bit younger than the one in 1975. Therefore, the capability of carbon sequestration of the northwestern forest ecosystem in 2000 was lower than that in 1975; the area acted as a carbon source. These results are in agreement with previous studies documenting that young forests generally emitted carbon to the atmosphere from the terrestrial biosphere due to higher ecosystem respiration than production (Pregitzer & Euskirchen, 2004; Magnani et al., 2007). This finding highlights the importance of land-use history in determining the regional carbon sink/source pattern.

7.4 Summary

Climate and land use and land cover change play important roles in determining the spatial and temporal dynamics of carbon stocks in terrestrial ecosystems and carbon fluxes between biosphere and atmosphere. Both annual precipitation and mean annual temperature increased in the Jinsha watershed from 1975 to 2000. Although there were not significant net changes in land covers at the regional level, various land covers did change at the finer scales. The Jinsha watershed ecosystem acted as a carbon sink during the period of 1975–2000, with an average rate of 0.36 Mg/ha/yr. Temporal variability of carbon dynamics was primarily attributed to regional climate change, while the spatial heterogeneity of ecosystem carbon sink/source pattern mainly resulted from local land use and land cover change. Carbon sinks were associated with forest areas with little disturbance, whereas carbon sources were caused by stand-replacing activities.

Our study has demonstrated the potential benefit of employing a systematic modeling approach to quantifying ecosystem carbon dynamics, or other processes and functions. To that end, a very promising venue is to integrate biophysical and biogeochemical models with geographic information systems by incorporating the driving forces at various spatiotemporal scales. Our study also implies that the forest protection and afforestation practices in the Jinsha watershed were conducive to terrestrial carbon storage, whereas the deforestation and land degradation activities were detrimental.

Acknowledgments This study was funded by the U.S. National Science Foundation (project #0507948). Logistical support from Sichuan Agricultural University is gratefully appreciated.

References

Achard, F., Eva, H. D., Mayaux, P., Stibig, H. J., & Belward, A. (2004). Improved estimates of net carbon emissions from land cover change in the tropics for the 1990s. *Global Biogeochemical Cycles, 18*, GB2008, doi:10.1029/2003GB002142

Bala, G., Caldeira, K., Wickett, M., Phillips, T. J., Lobell, D. B., Delire, C., et al. (2007). Combined climate and carbon-cycle effects of large-scale deforestation. *Proceedings of the National Academy of Sciences of the United States of America, 104*, 6550–6555

Binford, M. W., Gholz, H. L., Starr, G., & Martin, T. A. (2006). Regional carbon dynamics in the Southeastern US Coastal plain: Balancing land cover type, timber harvesting, fire, and environmental variation. *Journal of Geophysical Research-Atmospheres, 111*, D24S92, doi: 10.1029/2005JD006820

Canadell, J. G. (2002). Land use effects on terrestrial carbon sources and sinks. *Science in China Series C-Life Sciences, 45*, 1–9

Cao, M. K., Prince, S. D., Small, J., & Goetz, S. J. (2004). Remotely sensed interannual variations and trends in terrestrial net primary productivity 1981–2000. *Ecosystems, 7*, 233–242

Chapin, F. S., Woodwell, G. M., Randerson, J. T., Rastetter, E. B., Lovett, G. M., Baldocchi, D. D., et al.(2006). Reconciling carbon-cycle concepts, terminology, and methods. *Ecosystems, 9*, 1041–1050

Choi, S. D., Lee, K., & Chang, Y. S. (2002). Large rate of uptake of atmospheric carbon dioxide by planted forest biomass in Korea. *Global Biogeochemical Cycles, 16* 1089, doi:10.1029/2002GB001914

Conservation International. (2002). *Biodiversity hotspots*. Retrieved from http://www.biodiversityhotspots.org/xp/hotspots/China

Davidson, E. A., & Janssens, I. A. (2006). Temperature sensitivity of soil carbon decomposition and feedbacks to climate change. *Nature, 440*, 165–173

Editorial Committee for Panzhihua Statistical Yearbook (ECPSY) (2002). Panzhihua's statistical yearbook from 1987 to 2001. Statistical Bureau of Panzhihua, Sichuan

Fang, J. Y., Chen, A. P., Peng, C. H., Zhao, S. Q., & Ci, L. (2001). Changes in forest biomass carbon storage in China between 1949 and 1998. *Science, 292*, 2320–2322

Fang, J. Y., Piao, S. L., Tang, Z. Y., Peng, C. H., & Wei, J. (2001). Interannual variability in net primary production and precipitation. *Science, 293*, 1723a

Fang, J. Y., Piao, S. L., Field, C. B., Pan, Y. D., Guo, Q. H., Zhou, L. M., Peng, C. H., Tao, S. (2003). Increasing net primary production in China from 1982 to 1999. Frontiers in Ecology and the Environment, *1*, 293–297.

Fang, J. Y., Oikawa, T., Kato, T., Mo, W. H., & Wang, Z. H. (2005). Biomass carbon accumulation by Japan's forests from 1947 to 1995. *Global Biogeochemical Cycles, 19*, GB2004, doi:10.1029/2004GB002253

Foley, J. A., Defries, R., Asner, G. P., Barford, C., Bonan, G., Carpenter, S. R., et al. (2005). Global consequences of land use. *Science, 309*, 570–574

Gimona, A. Birnie, R. V., & Sibbald, A. R. (2006). Scaling up of a mechanistic dynamic model in a GIS environment to model temperate grassland production at the regional scale. *Grass and Forage Science, 61*, 315–331

Houghton, R. A. (2003). Revised estimates of the annual net flux of carbon to the atmosphere from changes in land use and land management 1850–2000. *Tellus B, 55,* 378–390

Houghton, R. A., & Goodale, C. L. (2004). Effects of land-use change on the carbon balance of terrestrial ecosystems. In DeFries, R. S., Asner, G. P., & Houghton, R. A. (Eds.), *Ecosystems and land use change*(pp. 85–98). Washington, DC: American Geophysical Union

Iverson, L. R., Dale, M. E., Scott, C. T., & Prasad, A. (1997). A GIS-derived integrated moisture index to predict forest composition and productivity of Ohio forests (USA). *Landscape Ecology, 12,* 331–348

Jenkins, J. C., Birdsey, R. A., & Pan, Y. (2001). Biomass and NPP estimation for the mid-Atlantic region (USA) using plot-level forest inventory data. *Ecological Applications, 11,* 1174–1193

Jenkinson, D. S., Adams, D. E., & Wild, A. (1991). Model estimates of CO_2 emissions from soil in response to global warming. *Nature, 351,* 304–306

Kauppi, P. E., Ausubel, J. H., Fang, J. Y., Mather, A. S., Sedjo, R. A., & Waggoner, P. E. (2006) Returning forests analyzed with the forest identity. *Proceedings of the National Academy of Sciences of the United States of America, 103,* 17574–17579

Liu, S. G. (2008). Quantifying the spatial details of carbon sequestration potential and performance. In B. McPherson & E. Sundquist (Eds.), *Science and technology of carbon sequestration.* Washington, DC: American Geophysical Union (in press)

Liu, S. G., Bliss, N., Sundquist, E., & Huntington, T. G. (2003). Modeling carbon dynamics in vegetation and soil under the impact of soil erosion and deposition. *Global Biogeochemical Cycles, 17,* 1074, doi:10.1029/2002GB002010

Liu, S. G., Kaire, M., Wood, E., Diallo, O., & Tieszen, L. L. (2004). Impacts of land use and climate change on carbon dynamics in South-Central Senegal. *Journal of Arid Environments, 59,* 583–604

Liu, S. G., Loveland, T. R., & Kurtz, R. M. (2004). Contemporary carbon dynamics in terrestrial ecosystems in the Southeastern plains of the United States. *Environmental Management, 33,* S442–S456

Lu, X. X. (2005). Spatial variability and temporal change of water discharge and sediment flux in the lower Jinsha tributary: Impact of environmental changes. *River Research and Applications, 21,* 229–243

Magnani, F., Mencuccini, M., Borghetti, M., Berbigier, P., Berninger, F., Delzon, S., et al. (2007). The human footprint in the carbon cycle of temperate and boreal forests. *Nature, 447,* 848–850

Meyer, W. B., & Turner, B. L. (1992). Human-population growth and global land-use cover change. *Annual Review of Ecology and Systematics, 23,* 39–61

Nemani, R. R., Keeling, C. D., Hashimoto, H., Jolly, W. M., Piper, S. C., Tucker, C. J., et al. (2003). Climate-driven increases in global terrestrial net primary production from 1982 to 1999. *Science, 300,* 1560–1563

Pan, Y. D., Melillo, J. M., Mcguire, A. D., Kicklighter, D. W., Pitelka, L. F., Hibbard, K., et al. (1998). Modeled responses of terrestrial ecosystems to elevated atmospheric CO_2: A comparison of simulations by the biogeochemistry models of the Vegetation/Ecosystem Modeling and Analysis Project (VEMAP). *Oecologia, 114,* 389–404

Potter, C. S., Randerson, J. T., Field, C. B., Matson, P. A., Vitousek, P. M., Mooney, H. A., et al. (1993). Terrestrial ecosystem production: A process model based on global satellite and surface data. *Global Biogeochemical Cycles, 7,* 811–842

Pregitzer, K. S., & Euskirchen, E. S. (2004). Carbon cycling and storage in world forests: Biome patterns related to forest age. *Global Change Biology, 10,* 2052–2077

Ramankutty, N., Gibbs, H. K., Achard, F., Defriess, R., Foley, J. A., & Houghton, R. A. (2007). Challenges to estimating carbon emissions from tropical deforestation. *Global Change Biology, 13,* 51–66

Rustad, L. E., Campbell, J. L., Marion, G. M., Norby, R. J., Mitchell, M. J., Hartley, A. E., et al. (2001). A meta-analysis of the response of soil respiration, net nitrogen mineralization, and aboveground plant growth to experimental ecosystem warming. *Oecologia, 126,* 543–562.

Schimel, D. S., House, J. I., Hibbard, K. A., Bousquet, P., Ciais, P., Peylin, P., et al. (2001). Recent patterns and mechanisms of carbon exchange by terrestrial ecosystems. *Nature, 414*, 169–172

Shi, X. Z., & Yu, D. S. (2002). A framework for the 1:1,000,000 soil database of China. *Proceedings of the 17th World Congress of Soil Science*, Bangkok

Tian, H. Q., Melillo, J. M., Kicklighter, D. W., Mcguire, A. D., Helfrich, J. V. K., Moore, B., et al. (1998). Effect of interannual climate variability on carbon storage in Amazonian ecosystems. *Nature, 396*, 664–667

Tickle, P. K., Coops, N. C., & Hafner, S. D. (2001). Assessing forest productivity at local scales across a native eucalypt forest using a process model, 3 pg-spatial. *Forest Ecology and Management, 152*, 275–291

Vitousek, P. M., Mooney, H. A., Lubchenco, J., & Melillo, J. M. (1997). Human domination of Earth's ecosystems. *Science, 277*, 494–499

Xiang, Q., Yin, R. S., Xu, J. T., & Deng, X. Z. (2009). Modeling the driving forces of land use and land cover changes along the upper Yangtze River of China. *Environmental Management*(under review)

Zhao, S. Q., Peng, C. H., Jiang, H., Tian, D. L., Lei, X. D., & Zhou, X. L. (2006). Land use change in Asia and the ecological consequences. *Ecological Research, 21*, 890–896

Chapter 8
Process-Based Soil Erosion Simulation on a Regional Scale: The Effect of Ecological Restoration in the Chinese Loess Plateau

Changbin Li, Jiaguo Qi, Zhaodong Feng, Runsheng Yin, Biyun Guo, and Feng Zhang

Abstract Land degradation caused by serious soil erosion has made the Loess Plateau one of the poorest regions in China. To improve the environmental conditions, the government has taken a number of measures of ecological restoration, including the Sloping Land Conversion Program (SLCP) since 1999. A natural question is whether these measures have improved the regional environmental conditions. The objective of this paper is to answer this question, focusing on the soil erosion dynamics in the Zuli River basin. We adopt a soil erosion model to simulate the changes of runoff and soil erosion induced by implementing the ecological restoration efforts, with the assistance of remote sensing and GIS technologies for parameterization of the land surface attributes. Our results show that the improved ground vegetative cover, especially in forestland and grassland, has reduced soil erosion by 38.8% from 1998 to 2006. At the same time, however, the changed rainfall pattern has resulted in a 13.1% increase in soil erosion. Thus the net reduction in soil erosion was 25.7%. This suggests that China's various ecological restoration efforts have been effective in reducing soil loss on a regional scale.

Keywords Ecological restoration · Soil erosion · Land use and land cover change · Precipitation pattern · Regional hydrological simulation

8.1 Introduction

China's demographic and economic growths have put great pressures on its environment and natural resources (Liu & Diamond, 2008). To reduce these pressures, the government has been undertaking a number of measures, including the huge Sloping Land Conversion Program (SLCP), also called Grain for Green Program

C. Li (✉)
Key Laboratory of Western China's Environmental Systems (Chinese Ministry of Education), Lanzhou University, Lanzhou, Gansu, 730000, Center for Global Change and Earth Observations, Michigan State University, East Lansing, MI 48823, USA

(Feng, Yang, Zhang, Zhang, & Li, 2005). The primary goal of the SLCP is to convert degraded and desertified farmland into forestland and grassland. Thus, a logical question is whether this program has improved the regional environmental conditions since its initiation in 1999. The objective of this paper is to address this question by examining the erosion dynamics before and after the SLCP implementation in the Zuli River basin (ZRB) of the Loess Plateau region of China.

Located in the middle reach of the Yellow River, the Loess Plateau region is notoriously known for its severe soil erosion. Over 60% of the land in the region suffers from various degrees of erosion as a consequence of unsustainable land use and vegetation degradation, as well as the intrinsic property of the deep, loose yellow soils found there. The regional erosion rate even reached 30,000 t km^{-2}yr^{-1} (Hessel, Jetten, Liu, Zhang, & Scolte, 2003; Li, Poesen, Yang, Fu, & Zhang, 2003; McVicar, Li, et al., 2007). This problem, along with other adverse factors such as water deficit and remote location, has in turn resulted in widespread rural poverty and made the regional sustainable development an elusive possibility (Shi & Shao, 2000). As such, the Loess Plateau region received the highest priority in implementing the SLCP, in an attempt to mitigate the worsening soil erosion problems by re-vegetation and afforestation of slope croplands. The effectiveness of the SLCP in this region has yet to be evaluated and the assessed outcome of erosion change can serve as a surrogate for the successfulness of the ecological restoration efforts over a ten-year period of implementation.

In this study, we adopt a distributed soil erosion model to simulate the changes of runoff and soil erosion across the Zuli River basin, induced by implementing the SLCP. We will do so with the assistance of the remote sensing (RS) technologies and the Geographic Information System (GIS) for parameterization of the land surface attributes. Our results show that the improved ground cover, especially forestland and grassland, has resulted in an erosion reduction of 38.8%, compared to the mean level of the 1990 s. On the other hand, the changed rainfall pattern has caused soil erosion to increase by 13.1%. In combination, we find a net decrease of soil erosion in the ZRB by 25.7% in a decade. This evidence suggests that the ecological restoration efforts have effectively mitigated the soil loss.

The chapter is organized as follows: We describe the study site in the next section and present our method and material in Section 8.3, then we report our simulation results in Section 8.4, and we make some closing remarks in the final section.

8.2 Study Site

With an area of 10,653 km^2, the ZRB is located in the upper reach of the Yellow River between 104°12′–105°33′E and 35°18′–36°34′N (Fig. 8.1). It features a hilly loess landscape and a semi-arid climate, with cold and dry winter months and warm and relatively wet summer months, as characterized by McVicar, Li, et al. (2007). Its elevation varies from 1,400 m in the north to over 2,800 m in the south. Accordingly, the mean annual precipitation ranges from over 500 mm in the south to less than 250 mm in the north, and the mean annual temperature varies from 3.6 to 8.8°C.

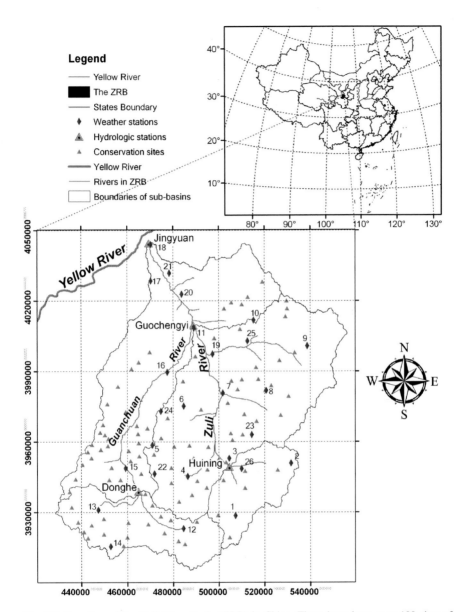

Fig. 8.1 Location of the Zuli River Basin (ZRB) in China. There have been over 100 sites of ecological restoration (*the light grey triangles*) across the ZRB, with almost half of them in the sub-basin of the River's main drainage area and others in the biggest tributary, the Guanchuan River. *Triangles with a black dot in them* are the 4 hydrologic stations where runoff and sediment are observed

The Zuli River itself is 219 km long and originates in the Huajia Mountain. Its longest tributary is the Guanchuan River in the west part of the basin. Driven by the regional land use and land cover changes (LULCC) and altered precipitation, among other things, the Zuli River's annual discharge has decreased from an average of 1.06×10^8 m^3 from 1970 to 1999 to less than 0.7×10^8 m^3 after 2000 (Li, 2006).

The intensive rainfall and the hilly terrain cause large amounts of overland flow during flood season, which is the primary natural driver for the high rate of soil erosion in the Loess Plateau region (Xu, 2005). In the ZRB, however, excessive removal of vegetation and cultivation-induced deterioration of the surface soil properties have exacerbated the problem. The ZRB once had the highest annual erosion yield among the six primary tributaries of over 10,000 km^2 in the upper reaches of the Yellow River, despite the fact that it is the smallest one among them (Li, 2006; Zhang, 2006).

It is difficult to identify any major tree species in the ZRB's hilly area due to the historical land clearance. The remaining vegetation is a mixture of native and introduced grass and shrub species, with a few trees planted around arable lands or along roads. To prevent the sediments from reaching the Yellow River, both re-vegetation (e.g., planting perennial grasses) and engineering methods (e.g., terracing and damming) are called for, which, of course, result in changes in land use and land cover. Changes in land use and land cover, in turn, affect the hydrological dynamics.

8.3 Method and Material

Many approaches have been used for water discharge and soil erosion simulation (e.g., Govindaraju, Reddi, & Kasavaraju, 1995; Felix-Henningsen, Morgan, Mushala, Richson, & Scholten, 1997; Morgan, 2001; Li et al., 2003; Rose et al., 2004; Vigiak, Sterk, Romanowicz, & Beven, 2006; Li, Feng, Ma, & Wei, 2008). Technological advances in earth observations and GIS have made it possible to extract useful information from satellite images and digital elevation models (DEM) for improved parameterization. For example, the Soil Erosion Model for Mediterranean regions is based on the method originally formulated by Morgan, Morgan, and Finney, or the MMF method for short (De Jong et al., 1999), which incorporates the spatial and temporal variations of such environmental factors as rainfall, soil capacity of moisture storage, regional topography, and cropping practices derived from remotely sensed imagery. In this study, the MMF method is adopted for an up-scaling application of the regional water-soil loss prediction with the advantages of its process-based conceptual erosion framework and its feature of scale matching from the site-based calibration method (see Appendix 1 for detail).

The data used include: (1) series of Landsat images (17 Jun., 1975; 27 Sept., 1978, 26 Aug., 1993, 21 Jun., 1998, 29 Jun., 2001, and 10 May, 2003), and MODIS EVI (Enhanced Vegetation Index) for 2006; (2) a DEM derived from a digital relief map at 1: 50,000 scale with the cell size of 100 m; (3) soil maps of China at 1:1,000,000 scale; (4) land use map in 1996 for calibration of the classification for

1998; (5) yearly hydro-meteorological data from 26 weather stations; and (6) runoff and sediment data series from four hydrologic stations. The weather and hydrological stations are shown in Fig. 8.1. All data, including the land cover information derived from the Landsat MSS, TM, ETM+ images, and MODIS product, were resampled to 100 m cell size, the same as the DEM data.

Table 8.1 lists the relevant parameters of the hydrologic properties for different soil types in the ZRB. Variables, such as annual rainfall and effective rainfall (Fig. 8.2a), were calculated from observations of the weather stations. Slope steepness (Fig. 8.2b), a variable that has spatial but not temporal variation, was computed from the 1:50,000 DEM. Plant/crop cover factor (Fig. 8.2c) was estimated based on Equation (8.11) in Appendix 1, and the normalized difference vegetation index (NDVI) was derived from satellite images (see Appendix 2). The percentage of ground cover (Fig. 8.2d) was empirically estimated based on the image classification. The effective hydrologic depth (Figs. 8.2e, f), a key factor controlling the land surface flow generation, was estimated on the basis of the semi-arid hydrological and soil properties in the Loess Plateau region.

The rainfall data from 1970 to 1989 were used for model calibration, and those for the next 10 years were used for model validation. The deterministic coefficient, also known as the coefficient of efficiency, was used for the model

Table 8.1 Soil parameter values

Definition	Texture	K (g/J)	MS (w/w)	BD (mg/cm^3)	COH (kPa)
Light sierozem	Sandy loam	0.7	0.21	1.2	2.00
Sierozem	Loamy sand	0.3	0.19	1.18	2.00
Salinized sierozem	Sandy loam	0.7	0.21	1.2	2.00
Cultivated black hemp soil in plain field	Sandy clay loam	0.1	0.29	1.28	3.00
Cultivated light black loessial soil in plain field	Sandy clay loam	0.1	0.29	1.3	3.00
Loessial soil in slope land	Silty loam	0.9	0.27	1.19	3.00
Cultivated light sierozem in sand farmland	Sandy clay loam	0.1	0.29	1.34	3.00
Loessial soil	Silt	1.0	0.25	1.17	2.50
Cultivated and irrigated light sierozem in plain field	Sandy clay loam	0.1	0.29	1.31	3.00

K, MS, BD and COH were defined and parameterized according to Morgan (2001), with empirical modifications based on field surveys and experiments.

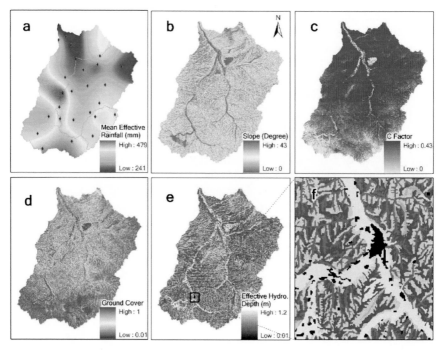

Fig. 8.2 Spatial Distributions of Some Important Variables and Parameters. (**a**) Multi-Year Average Precipitation-Based Effective Rainfall in the ZRB. (**b**) Slope Steepness Across the ZRB. (**c**) The C Factor Derived from NDVI – the Higher the Latter, the Lower the Former. (**d**) Ground Cover Defined by Coupling Field Surveys with Image Classification. High Ground Cover Values Often Correspond to Low C Factors Spatially and Less Soil Loss. (**e**) Effective Hydrological Depth of Soil Varies Spatially, Depending on Land Cover. It Resulted in More Acceptable Calibration Outputs. (**f**) Local Amplification Map of Parameter EHD. Urban Area with the Solid Surface Has Higher Overland Runoff Ratio, So EHD There Is Near Zero (*Black Region*); Valleys with Much Plain Farmland Had a Deeper Range of EHD Because of the Greater Infiltration, Leading to a Lower Overland Flow Generation (*Light Grey Region*). *Solid Diamonds* Represent the Weather Stations with Their Names and Serial Numbers in Fig. 8.2a, b, Respectively

accuracy assessment. The deterministic coefficient has been widely used to evaluate the performance of hydrologic models (Nash & Sutcliffe, 1970; Bekele & Nicklow, 2007). That is,

$$c_e = 1 - \frac{\sum_{j=1}^{n} (O_j - S_j)^2}{\sum_{j=1}^{n} (O_j - \overline{O})^2} \qquad (8.1)$$

where c_e is the coefficient of efficiency, O_j and S_j are observed and simulated basin response (e.g., discharge and sediment concentration), $\overline{O}$ is the mean of observed basin response, n is the length of the data series. Based on the spatial distribution of the calculated annual discharge and sediment concentration, the average

Table 8.2 Accuracy assessment of simulations based on the MMF method

Hydrologic stations	Catchments area (km^2)	Deterministic coefficient	
		Discharge	Sediment
Jingyuan	10,653	0.8540	0.7443
Guocheng	5,462	0.7387	0.6387
Huining	1,007	0.5275	0.5860
Chankou	1,645	0.7423	0.7012
Average	–	0.7156	0.6675

deterministic coefficient of our simulations is 0.72 for runoff and 0.67 for sediment (Table 8.2). These figures indicate a fairly high accuracy of our regional hydrologic and erosion simulations.

8.4 Simulation Results

Simulations were made in correspondence to the land use types under scenarios of the variable and constant annual rainfalls to allow us to distinguish the impacts of precipitation from those of land use change on the hydrologic and soil processes. The ground cover fraction was derived from the basis of classified land use types in different periods of time. It was found that, relative to the average rate of the 1990s, the basin discharge reduced by 3.6–35.3% during the period of 1999–2006, even though the annual precipitation fluctuated from −15.7 to 17.5%.

8.4.1 Land Use and Land Cover Changes

The land use classification from remotely sensed imagery was deemed to be fairly good, given the overall accuracy rate of 77.92% and the Kappa statistic of 0.6901 (Table 8.3). Overall, cropland, grassland, and unused land are the three dominant categories of land use and they account for over 90% of the land area altogether. In contrast, the proportions of forestland and other land (bodies of water and built-up land) are small; also, their changes show irregular patterns. These irregularities resulted from errors caused by the image spatial resolutions and the selection of training sites for classification. In particular, when MSS and MODIS images were processed, the small proportions of water and built-up areas could not be clearly identified due to their coarse spatial resolution.

The cropland area was 43.5 × 10^4 ha (occupying 40.7% of the basin area) in 1970, and increased to 50.1 × 10^4 ha (46.9%) by the early 1990 s. During the same time period, grassland increased from 23.3 × 10^4 ha (21.8%) to 25.8 × 10^4 ha (24.2%). From 1998, however, cropland began to decline while grassland continued

Table 8.3 Accuracy of image classification

Year	Data	Cell size (m)	Overall classification accuracy (%)	Overall Kappa statistics
1970	MSS	57	87.88	0.8261
1993	TM	28.5	75.00	0.6439
1998	TM	28.5	85.29	0.8117
2001	ETM+	28.5	78.05	0.7064
2003	ETM+	28.5	77.42	0.6506
2006	MODIS	250	63.89	0.5021
Average			77.92	0.6901

increasing. During the period of 1998–2006, the proportion of cropland decreased from 48.6 to 37.2%, whereas grassland increased from 22.9 to 26.6%. Another associated trend was the tremendous decrease of unused land from 35.8% in 1970 to 17.3% in 2001; thereafter, however, it increased to 28.6% by 2006. Throughout the 1970 s, 1980 s, and 1990 s, forestland kept expanding slowly, and its coverage has recently reached 8% of the ZRB (Fig. 8.3). In sum, the area of grassland and forestland increased from 27.4% (4.5% + 22.9%) in 1998 to 34.1% (7.5% + 26.6%) in 2006.

Undoubtedly, the positive land-use trends since the late 1990 s have had much to do with implementing the Grain for Green Program. In this context, the large gain of unused land can be understood as an increase in abandoned cropland, on which efforts had been made to plant trees or to establish grass cover. Nevertheless, the trees and grass cover were still not detectable in the remotely sensed imageries due to their slow growth, caused by the limited water availability (McVicar, Li, et al., 2007).

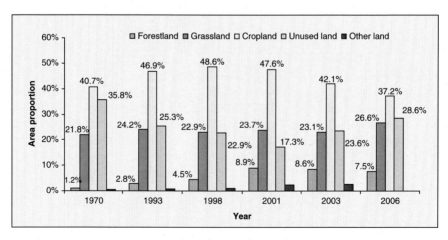

Fig. 8.3 Changes of the Land Use Types from 1970 to 2006

8.4.2 Soil Erosion Change with Land Uses

To examine the influence of a single type of land use on soil erosion, simulations were run with annual rainfall data for each land use type found in the region, to obtain the runoff and sediment for each land cover type. Further, treating the value from cropland simulation as a reference, ratios of runoff and sediment of other land use types were calculated. For unused land, the runoff ratio was 5.33, while the sediment ratio was 7.60. For grassland and forestland, however, the ratios were quite low, being only 0.68 and 0.15 for grassland and 0.59 and 0.11 for forestland (Table 8.4).

Table 8.4 Average water-sand status under different land uses

Items	Land use	Value	Ratio
Runoff rate	Farmland	28.75	1.00
(m^3/ha/a)	Unused land	153.29*	5.33
	Grassland	19.67	0.68
	Forest land	4.35	0.15
Sediment	Farmland	12.99	1.00
rate	Unused land	98.72*	7.60
(t/ha/a)	Grassland	7.64	0.59
	Forest land	1.39	0.11

*The high runoff and erosion rate of unused land are mainly because of the extremely sparse land cover and the loose property of loess.

There were two major causes of the relative high runoff and erosion rates of unused land. First, as noted earlier, the vegetation cover of unused land is low (typically less than 10%), as detected on satellite imagery. Therefore, it is not surprising to see high runoff and erosion, for vegetative cover is one of the most important factors in soil loss calculations. On the other hand, water absorption in fields of annual staple crops played a major role in reducing soil erosion. And rainfall on other lands cultivated for cash crops, most of which tend to be in flatter and less compacted areas, may completely infiltrate to deeper soils if the precipitation intensity is low. Field evidence showed that there is little overland flow or erosion when the rainfall intensity was below 8.0 mm/h (Soil Conservation Service of Gansu Provincial Water Resources Bureau, 1989). These differences suggest that changing land uses in the ZRB, as shown in Table 8.3, can potentially reduce soil erosion. That is, either change in land use or improvement in vegetative cover in the ZRB, which is a primary goal of the ecological restoration projects, may lead to significant reduction of soil erosion.

8.4.3 Spatial Characteristics of Soil Erosion

The large spatial variations of precipitation and heterogeneous landscape resulted in sharp differences in soil erosion across the ZRB. Vegetative covers in the upper

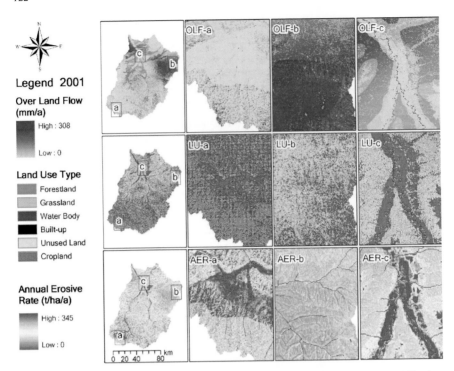

Fig. 8.4 Spatial Characteristics of Land Uses and Simulated Overland Flow and Soil Erosion. Area a, b, and c Are Selected Regions for Spatial Comparisons of Soil Erosion from the Upper to the Lower Reaches. Over Land Flow and Annual Erosion Rate, the Most Important Outputs from the MMF Model, are Computed from Equations (8.4) and (8.12) for 2001 and Arranged in the Row Order of a, b, and c in Fig. 8.5. Land Use Type is the Significant Factor in Both Temporal and Spatial Variation of Soil Erosion

reach of the basin were higher than those in the middle and lower reaches, leading to less soil erosion.

Figure 8.4 depicts the distribution of the simulated overland flow (OLF, in mm), the corresponding land use types (LU), and the computed annual erosion rate (AER, in kg/ha) for 2001. Better vegetation conditions in the upper reaches, with more forest and less unused land, correspond to lower runoff rate and thus smaller erosion yield. The fraction of forest and grass cover (LU-a) is obviously higher than that of unused land with sparse vegetation in an intensely eroded region in the eastern ZRB (LU-b), or that of the cropland-dominated landscape near a river channel in the middle-lower reaches (LU-c).

In comparison, there is much more effective rainfall in the south and less in the drier north and east (Fig. 8.2a). However, because the overland flows (OLF-a, OLF-b and OLF-C) and the annual erosion rates are inversely related to the land cover conditions (LU-a, LU-b, LU-c), more intense soil erosion is found in the unused land, cropland, and sparse grassland-dominated region (AER-b) than in the forest, dense grassland, and cropland-dominated region (AER-a). In the cropland-dominated areas near a river channel, the erosion yield is low (the red areas in AER-c) since much of the rainfall is intercepted by the vegetation and the land is relatively flat.

Based on the geo-morphological routing algorithm in the GIS framework, the cumulative flow in river channels (with a threshold area of 10 km^2) can be computed as well. Higher cumulative channel flow leads to higher flow detachment and channel transport rates, causing a much greater annual erosion rate in the channels. Thus, erosion from the channel area constitutes a large fraction of the total sediment of the ZRB (blue bands in AER-a, b and c).

8.4.4 Impacts of Land Use Change on Soil Erosion

To assess the LULCC impact at the regional scale, temporally distributed effective rainfall was used as input in simulating water-soil dynamics. Figure 8.5shows how the rainfall trend (light blue bars) affected soil erosion (pink line with triangle spots) and how both rainfall trend and land use changes (green line with round spots) affected soil erosion. The two colored straight lines are trends of the rainfall effect and the combined effect of rainfall and land use change since 1998, when the Grain for Green Program began. Against the hypothetical scenario of unchanged land use and constant rainfall (black line in Fig. 8.5), the rainfall trend exacerbated the erosion process by 13.1%, while the combined land use and rainfall changes mitigated the erosion by 25.7%. The total impact of vegetation restoration on erosion reduction – the difference between the above two values – is approximately

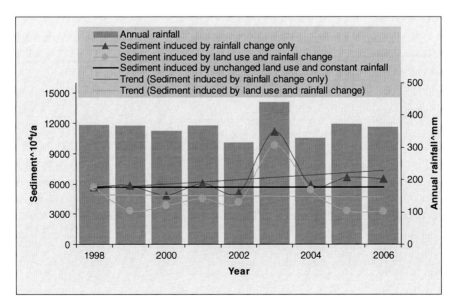

Fig. 8.5 Impacts of Rainfall and Land Use Changes on Soil Erosion

Fig. 8.6 Main Forms of Converting Corpland into Forest and Grassland Cover in the ZRB

38.8%. Put differently, the ecological restoration projects, including the Grain for Green Program, have helped mitigate soil erosion by 38.8% over the past 10 years (Fig. 8.6).

8.5 Conclusions and Discussion

In this study, a distributed soil erosion model was calibrated for a regional erosion assessment within the Zuli River Basin of the Loess Plateau in China. The model was run with different land uses and precipitation patterns, in an effort to quantify the potential impact of the regional ecological restoration efforts, including the Grain for Green Program. Our results indicate that in the semi-arid ZRB, sediment is driven primarily by the anti-erodibility of the soils and the rainfall patterns. Corresponding to the implementation of the SLCP, there have been major land cover changes across the ZRB during the period 1999–2006, as reflected mainly in the increase of forestland and grassland from 27.4 to 34.2% and the decrease of cropland from 48.6 to 37.2%. Meanwhile, compared to the mean level of the 1990 s, the annual rainfall varied from –15.7 to 17.5%. In combination, these cumulative LULCC and rainfall changes resulted in a reduction of overland flow from 3.6 to 35.3%. While the fluctuating rainfall pattern caused an increase in erosion by 13.1%, the land use changes led to an erosion reduction of 38.8%. The overall erosion reduction was thus estimated to be 25.7%. Now, it can be stated that the regional ecological restoration efforts have effectively mitigated the soil loss and thus contributed to the improvement of ecological conditions.

As interesting and informative as our results may be, caution should be taken in interpreting them. Given the large spatial scale and the process-based approach of our study on the one hand and the difficulty of gathering all of the required data and parameters on the other, what we have reported in this paper remains preliminary. In the future, we plan to devote further efforts to refining our modeling approach while deriving more accurate and longer data series and more representative parameters. Of course, this implies that a long-term monitoring project must be conducted (McVicar, Li, et al., 2007). To that end, we call for active participation and collaboration by the relevant government agencies and academic institutions in the region.

It is also worth noting that other restoration efforts prior to the SLCP should have helped improve the environmental conditions as well, even though substantial changes in regional land use and land cover began in the late 1990 s after the SLCP was initiated. As such, it would be inappropriate to attribute the reduced erosion solely to implementing the SLCP. Moreover, our findings highlight the fact that what is central to mitigating erosion is improved ground cover, rather than land retirement itself. To improve ground cover, satisfactory survival and stocking rates are crucial. In this regard, opportunities abound for enhancing the implementation of the SLCP. During our site visits, we found that cropland plots located in the relatively remote and inaccessible hilly areas or on the steep slopes were mostly converted to shrub or grass coverage (Li, 2006). As a result, the original loose top layer of the loess surface had already changed into a more stable status with an enhanced erosion-resistant capability. On the other hand, the planted trees and grass had very low survival and growth rates due to water deficit in quite a few of the so-called demonstration sites near urban centers and main access roads, leading to the scattered presence of "little old-man trees" (McVicar, Li, et al., 2007). In addition, where the land conversion was carried out in conjunction with the formation of such water-collecting and runoff-preventing structures as contour ditches and half-circle pits, the overland drainage and soil erosion were well contained. Similarly, where the established vegetation cover could be irrigated with stored rainwater, the ground hydrological and erosive processes would be modified favorably.

In short, the regional authorities and local communities have a lot to gain from a prompt and continuous assessment of the successes and failures in their past restoration efforts and from incorporating appropriate biological and engineering measures into their actions. It seems that one of the key tasks at this time is to swiftly and significantly improve the vegetation cover on those plots of unused land. Also, the program managers and researchers should be aware of the delicate water balance in the region. Ecological restoration, while controlling soil erosion in the upper and middle streams of the Yellow River, may substantially reduce water flows downstream in the northern plains, which can have detrimental consequences (McVicar, Van Niel, et al., 2007). Therefore, any effort assessing the effectiveness and impact of a restoration program must be done at different scales so as to systematically understand all of the positive and negative changes that are induced by the program.

Acknowledgements This work was supported by the China NSF Innovation Team Project (40721061), Chinese Academy of Science's grant "Ecosystem Services and Human Activity" at the Institute of Geographic Sciences and Natural Resources Research (IGSNRR), the grant by NASA's LCLUC program (NNG05GD49G) and the US NSF grants (0507948 and 0624018). The authors are grateful for Erin Shi and Victoria Hoelzer-Maddox for their assistance.

Appendix 1: The MMF Method

The MMF method for predicting annual soil erosion begins by dividing the study area into a number of Hydrological Response Units (HRUs) according to the land cover, soil type, and slope. In an HRU, the hydrological factors that influence the soil erosion process are homogeneous, with mean values of its variables as inputs. The simulation is carried out at a cell size of 100 m, which suits the resolution requirement for process-based environmental modeling at the regional scale (Li et al., 2008).

The erosion process is divided into soil particle detachment by raindrop impact and solid mass transport by runoff (Morgan, Morgan, & Finney, 1984; De Jong et al., 1999, Morgan, 2001). The annual erosion rate (AER, kg/m^2) is determined by the lesser of annual detachment rate (ADR, kg/m^2) and annual transport capacity (ATC, kg/m^2).

After modification (Morgan, 2001), the MMF procedure for calculating soil particle detachment by raindrop energy explicitly takes into account plant canopy height and leaf drainage. Tested with new field data, it performed well, as indicated by the high values of the coefficient of efficiency (CE) from a reduced regression between the observed and predicted values. Extensive applications also indicated that it can yield reasonable predictions under a wide range of conditions – from temperate climates to the Mediterranean climate and tropical monsoon climate regions Morgan (1985). The MMF method comprises a water phase and a sediment phase (Morgan et al., 1984). Soil particle detachment by raindrop impact occurs during the water phase, whereas erosion occurs during the sediment phase.

Water Phase Modeling

Energy of raindrop impacting soil detachment comes from two sources. One is the effective rainfall (ER, mm) that reaches the ground during the precipitation, falling between 0 and 1 after allowing for rainfall interception. The other is the water quantity intercepted by plant canopy and reaching the ground as leaf drainage (LD, mm). Because the ZRB features a semi-arid climate, the canopy interception can be considered as exhausted by post-rainfall transpiration; as such, the energetic impact from LD will not be included when predicting the regional erosion.

Then, the kinetic energy (KE, J/m^2) of the direct raindrop can be determined by a typical value for intensity of erosive rain (I, mm h^{-1}):

$$KE = ER(9.81 + 10.60 \log I) \tag{8.2}$$

There are several formulas calibrated worldwide for the relationship between KE and I. Formula (8.2) was developed by Onaga et al. for Okinawa, Japan, and it can be used across Eastern Asia (Onaga, Shirai, & Yoshimaga, 1998; Morgan, 2001).

Then, soil particle detachment by raindrop impact (DR, kg/m^2) can be simplified as:

$$DR = K \times KE \times 10^{-3} \tag{8.3}$$

where K is the erodibility of the soil (g/J) (Morgan, 2001). Parameters for K exist in correspondence to a wide range of soil textures (Table 8.1).

As a prerequisite, overland flow is crucial for computing the soil particle detachment by runoff. The procedure for estimating annual runoff (Q, mm) is based on the approach proposed by Kirkby (1976), which is based on the excessive generation theory when the daily rainfall exceeds the soil moisture storage capacity (R_c, mm). The annual runoff is obtained from:

$$Q = RG \times \exp(-R_c/R_0) \tag{8.4}$$

where RG (mm) is the ground-reaching portion of the rainfall; R_0 is the mean rainfall per rainy day (mm/d), i.e., RG/n, where n is the number of the rainy days in a specific year.

Quantifying the plant interception is important for calculating RG. The proportion of canopy interception to annual total rainfall in the ZRB is calibrated based on monthly observations, with a value of 3.201% (Li, 2006). This value is used for calculating the annual RG:

$$RG = R \times (1 - 0.03201) \tag{8.5}$$

Then, the soil capacity of moisture storage is estimated as:

$$R_c = 1000MS \times BD \times EHD \times (E_t/E_0) \tag{8.6}$$

where MS is the soil moisture content at field capacity (%, w/w); BD is the soil bulk density (mg/cm^3); EHD (m) is the effective hydrological depth of the soil. E_t/E_0 is the ratio of actual to potential evapotranspiration. While parameter values for MS, BD, and EHD can be found in Morgan (2001), they have been modified according to the local specificity and field experiments in this study.

Sediment Phase Modeling

The soil particle detachment by overland flow (runoff) (DO, kg/m^2) can then be determined by considering it as a function of runoff (Q), slope steepness (S, degree) and the soil resistance to erosion (Z):

$$DO = Z \times Q^{1.5} \times \sin S \times (1 - GC) \times 10^{-3} \qquad (8.7)$$

where GC is the percentage of ground cover. Equation (8.7) assumes that soil particle detachment by overland flow occurs only where the soil is not protected by ground cover. Z is a key parameter for determining the ground cover status. For loose, non-cohesive soils, $Z = 1$. Morgan (2001) emphasized soil cohesion (COH, kPa) as a primary component of its resistance to erosion:

$$Z = 1/(0.5COH) \qquad (8.8)$$

As listed in Table 8.1, values of COH can be obtained by matching soil types to those used in the European Soil Erosion Model (Morgan, Quinton, & Rickson, 1993).

The annual detachment rate (ADR) is the sum of the soil particles detachment caused by raindrop (DR, kg/m^2) and overland flow (DO, kg/m^2):

$$ADR = DR + DO \qquad (8.9)$$

The formula for estimating the annual transport capacity of the runoff (ATC) is:

$$ATC = C \times Q^2 \times \sin S \times 10^{-3} \qquad (8.10)$$

where C (0–1) is a plant or crop cover factor that can be adjusted to account for different tillage practices and levels of crop residue retention, and it can be derived from satellite imagery (De Jong, 1994; Knijff, Jones, & Montanarella, 1999):

$$C = 0.431 - 0.805 \times NDVI \qquad (8.11)$$

Finally, the smaller value of ADR and ATC is assigned as the annual erosion rate (AER):

$$AER = Min(ADR, ATC) \qquad (8.12)$$

Appendix 2: Imagery Processing for Estimating Land-Cover Changes

Remote sensing imagery can be used for detecting land surface features, such as cover status and vegetation impoverishment – key variables for environmental modeling McVicar & Jupp (1998). In applying the MMF method in this study, the percentage of ground cover (GC) and plant/crop cover factor (C) are derived from interpreting satellite images, based on the supervised classification technique and the vegetation index NDVI.

Supervised Classification for the LULCC

Supervised classification is done by training the computer system to recognize the land use patterns based on the signature catalogs. Images are first classified into more than 10 subclasses that are then merged into five main classes – forestland, grassland, cropland, unused land, and other land (built-up land and bodies of water, see Fig. 8.3). In this study, the Landsat 2 and Landsat 3 Multispectral Scanner (MMS) images from the 1970 s are used for mapping the land cover mosaic; those from the 1980 s, 1990 s, and 2000 s are derived using Landsat 5 and Landsat 7 Thematic Mapper (TM) images; and a map of MODIS (Moderate Resolution Imaging Spectroradiometer) product, EVI (Enhanced Vegetation Index), is adopted for 2006. Interpreting the images in terms of the classification system, we can determine the area of each type of land use.

NDVI for the factor C

In this study, the normalized difference vegetation index (NDVI), defined as the ratio of the difference and sum of the infrared (4th) and the red (3rd) bands by their digital numbers (DN), is calculated for determining the plant/crop cover factor used in Equation (8.11) for each period of time:

$$NDVI = (Band4 - Band3)/(Band4 + Band3) \tag{8.13}$$

References

Bekele, E. G., & Nicklow, J. W. (2007). Multi-objective automatic calibration of SWAT using NSGA-II. *Journal of Hydrology, 341*, 165–176.

De Jong, S. M. (1994). Vegetation parameters derived from satellite images for erosion modelling. *Earth Surface and Processes and Landforms, 19*, 1–14.

De Jong, S. M., Paracchini, M. L., Bertolo, F., Folving, S., Megier, J., & De Roo, A. P. J. (1999). Regional assessment of soil erosion using the distributed model SEMMED and remotely sensed data. *Catena, 37*, 291–308.

Felix-Henningsen, P., Morgan, R. P. C., Mushala, H. M., Richson, R. J., & Scholten, T. (1997). Soil erosion in Swaziland: A synthesis. *Soil Technology, 11*(3), 319–329.

Feng, Z., Yang, Y., Zhang, Y., Zhang, P., & Li, Y. (2005). Grain-for-green policy and its impacts on grain supply in West China. *Land Use Policy, 22*(4), 301–312.

Govindaraju, R. S., Reddi, L. N., & Kasavaraju, S. K. (1995). A physically based model for mobilization of kaolinite particles under hydraulic gradients. *Journal of Hydrology, 172*, 331–350.

Hessel, R., Jetten, V., Liu, B., Zhang, Y., & Scolte, J. (2003). Calibration of the LISEM model for a small Loess Plateau catchment. *Catena, 54*(1–2), 235–254.

Kirkby, M. J. (1976). Hydrological slope models: The influence of climate. In E. Derbyshire (Ed.), *Geomorphology and climate* (pp. 247–267). London: Wiley.

Knijff, J. M. V. D., Jones, R. J. A., & Montanarella, L. (1999). *Soil erosion risk assessment in Italy*. Space Application Institute, European Soil Bureau, pp. 25–26.

Li, C.-B. (2006). *Research of distributed hydrologic modeling in a typical large scale watershed in Longxi Loess plateau of China* (Ph.D. Dissertation, Lanzhou University). pp. 52–56 (In Chinese).

Li, C.-B., Feng, Z.-D., Ma, J.-Z., & Wei, G.-X. (2008). Study on spatiotemporal resolution for the regional distributed hydrological simulation. *Arid Zone Research (in Chinese), 25*(2), 169–173.

Li, Y., Poesen, J., Yang, J., Fu, B., & Zhang, J. (2003). Evaluating gully erosion using 137Cs and 210Pb/137Cs ratio in a reservoir catchment. *Soil and Tillage Research 69* (1–2), 107–115.

Liu, J., & Diamond, J. (2008). Revolutionizing China's environmental protection. *Science, 319*, 37–38.

McVicar, T. R., & Jupp, D. L. B. (1998). The current and potential operational uses of remote sensing to aid decisions on Drought Exceptional Circumstances in Australia: A review. *Agricultural Systems, 57*, 399–468.

McVicar, T. R., Li, L., Van Niel, T. G., Zhang, L., Li, R., Yang, Q. K., et al. (2007). Developing a decision support tool for China's re-vegetation program: Simulating regional impacts of afforestation on average annual streamflow in the Loess Plateau. *Forest Ecology and Management, 251*(1–2), 65–81.

McVicar, T. R., Van Niel, T. G., Li, L. T., Hutchinson, M. F., Mu, X. M. and Liu, Z. H. (2007). Spatially distributing monthly reference evapotranspiration and pan evaporation considering topographic influences. *Journal of Hydrology, 338*, 196–220.

Morgan, R. P. C. (1985). The impact of recreation on mountain soils: towards a predictive model for soil erosion. In: Bayfield, N. G., Barrow, G. C. (Eds.), The ecological impacts of outdoor recreation on mountain areas in Europe and North America. *Rural Ecology Research Group Report 9*: 112–121.

Morgan, R. P. C. (2001). A simple approach to soil loss prediction: A revised Morgan-Morgan-Finney model. *Catena, 44*(4), 305–322.

Morgan, R. P. C., Morgan, D. D. V., & Finney, H. J. (1984). A predictive model for the assessment of soil erosion risk. *Journal of Agricultural Engineering Research, 30*, 245–253.

Morgan, R. P. C., Quinton, J. N., & Rickson, R. J. (1993). *EUROSEM: A user guide.* Silsoe: Silsoe College, Cranfield University.

Nash, J. E., & Sutcliffe, J. V. (1970) River flow forecasting through conceptual models part I – A discussion of principles. *Journal of Hydrology, 10*(3), 282–290.

Onaga, K., Shirai, K., & Yoshimaga, A. (1998). Rainfall erosion and how to control its efforts on farmland in Okinawa. In S. Rimwanich (Ed.), *Land conservation for future generations* (pp. 627–639). Bangkok: Department of Land Development.

Rose, C. W., Gao, B., Walter, M. T., Steenhuis, T. S., Parlange, J.-Y., Nakano, K., et al. (2004). Reply to comment on "Investigating ponding depth and soil detachability for a mechanistic erosion model using a simple experiment by Gao, B., et al., (2003) Journal of Hydrology 277, 116–124". *Journal of Hydrology, 289*(1–4), 307–308.

Shi, H., & Shao, M. (2000). Soil and water loss from the Loess Plateau in China. *Journal of Arid Environments, 45*(1), 9–20.

Soil Conservation Service of Gansu Provincial Water Resources Bureau. (1989). Influences of water and soil conservation to hydrological and erosive processes across the Zuli River Basin (in Chinese).

Vigiak, O., Sterk, G., Romanowicz, R. J., & Beven, K. J. (2006). A semi-empirical model to assess uncertainty of spatial patterns of erosion. *Catena, 66*(3), 198–210.

Xu, J. (2005). The water fluxes of the Yellow River to the Sea in the past 50 years, in response to climate change and human activities. *Environment Management, 35*(5), 620–631.

Zhang, X. (2006). *GIS assisted modeling of soil erosion and hydro-geomorphic processes in the Zuli River Basin.* Ph. D Dissertation, Lanzhou University, Lanzhou, pp 25–32 (In Chinese).

Chapter 9
Conservation Payments, Liquidity Constraints and Off-Farm Labor: Impact of the *Grain for Green* Program on Rural Households in China

Emi Uchida, Scott Rozelle, and Jintao Xu

Abstract This study evaluates the labor response of rural households participating in the *Grain for Green* program in China, the largest payments for ecosystem services program in the developing world. Using a panel data set that we designed and implemented, we find that the participating households are increasingly shifting their labor endowment from on-farm work to the off-farm labor market. However, the effects vary depending on the initial level of human and physical capital. The results support the view that one reason why the participants are more likely to find off-farm employment is because the program is relaxing households' liquidity constraints.

Keywords China · *Grain for Green* program · Off-farm labor supply · Payments for ecosystem services · Program evaluation

9.1 Introduction

In the past decade, an increasing number of incentive-based conservation programs have been launched in the economies of developing countries, including Costa Rica, Mexico and China (e.g., Pagiola, Landell-Mills, & Bishop, 2002; Alix-Garcia, de Janvry, & Sadoulet, 2003; Hyde, Belcher, & Xu, 2003; Mayrand & Paquin, 2004). Often called payments for ecosystem service (PES), these incentive-based programs provide financial incentives to those who "supply" ecosystem services, including farmers who agree to set aside sensitive land or adopt farming technologies that generate ecosystem services such as wildlife habitat protection, carbon sequestration and protection of watershed functions.

E. Uchida (✉)
Department of Environmental and Natural Resource Economics, University of Rhode Island, Kingston, RI 02881, USA
e-mail: emi@uri.edu

R. Yin (ed.), *An Integrated Assessment of China's Ecological Restoration Programs*,
DOI 10.1007/978-90-481-2655-2_9, © Springer Science+Business Media B.V. 2009

Since farmers often are suppliers of these ecosystem services, programs often have been designed with dual goals—to generate ecosystem services *and* to achieve rural development (Pagiola & Platais, 2005). A PES program can increase farmers' income directly and indirectly through payments. For example, farmers who agree to set aside previously cultivated land for conservation purposes can increase their incomes if the payments exceed the opportunity cost of land retirement. Farmers can also use the payments to finance other productive activities, both on and off the farm. Depending on the program design, these schemes can induce a reallocation of factor endowments, and therefore indirectly induce structural changes in household income earning activities.

The programs may be unsuccessful, however, if they cannot induce farmers to transform their income generating activities. Payments are typically made for only a fixed term and can be terminated early due to political disagreements and/or budget constraints. In the longer run, farmers often must shift their agricultural practices and income generating activities so that they do not rely on program compensation payments. Otherwise, upon their termination of the activities, farmers may have to return the land to cultivation to survive, undoing the program's long-term environmental benefits.

Despite the importance of understanding how farmers change their labor allocation patterns in response to these programs, few studies to date have examined how PES schemes have affected farmers' allocation of factor endowments or choice of income generating activities. Many critical questions remain. For example, how does a conservation set-aside program induce farmers to shift labor allocations from on-farm production to off-farm work? What is the effect of such programs on on-farm labor allocation? Do program impacts depend on the participants' physical and human capital endowments?

This study examines these questions by analyzing the largest PES experiment in the developing world: the *Grain for Green* program in China. Following a series of devastating floods in 1998, China's government initiated a conservation set-aside program known as *Grain for Green*.[1] When the community is chosen to be part of the program, households can choose (or are allowed to choose) to set aside all or part of the cultivated land on sloped cropland and plant them with tree seedlings. As we elaborate later on, however, most observers know that *Grain for Green* has been "quasi voluntary."

The program's first objective is to increase forest cover on sloped cultivated land in the upper reaches of the Yangtze and Yellow River basins to prevent soil erosion. Its second objective is to alleviate poverty and restructure agricultural production into more environmentally and economically sustainable activities in some of the poorest parts of rural China (State Forestry Administration, 2003). According to interviews that we have conducted over the past several years, local governments

---

[1] The program was officially implemented in 2001. Pilot projects for the program got under way in 1999 in selected provinces. The *Grain for Green* program is also known as the Sloped Land Conversion Program.

consider the program as an opportunity to promote transformation of their counties' local economic structures. A survey of investment projects between 1998 and 2003 in 2,459 sample villages across six provinces in China showed that this program was the third most common project being implemented after roads and irrigation projects (Zhang, Luo, Liu, & Rozelle, 2006).

More than 5 years into the program, however, it is not yet clear how *Grain for Green* has affected how farmers allocate labor across income generating activities. On one hand, the government clearly expects that the program will facilitate a shift in labor from low-profit grain production to the production of more profitable crops and livestock and, more importantly, from on-farm work to off-farm work. On the other hand, off-farm activities, including self-employment and wage income earning activities, both in local job markets and in migrant labor markets, have been a driving force in reducing poverty in rural China (deBrauw, 2002; Bowlus & Sicular, 2003). Given this recent trend, households in rural China are likely to be increasing off-farm activities even when they were not enrolled in *Grain for Green*. Empirical findings on the program's labor impact are mixed: two studies used data collected 2 years after the program began and found that the program had no impact on off-farm employment (Xu, Bennett, Tao, & Xu, 2004; Uchida, Xu, Xu, & Rozelle, 2007). A study involving data collected 4 years into the program found a positive effect on off-farm labor participation (Groom, Grosjean, Kontoleon, & Swanson, 2006).[2]

In fact, this study of the impact of *Grain for Green* on labor allocation in China is part of a wider set of studies examining the fundamental question of how government payments affect the off-farm labor decisions of farmers, a subject of long interest to agricultural economists. During the past three decades, off-farm activities have provided a critical income source to a majority of farm households in the US and off-farm provision has been largely responsible for closing the gap in income between farm and nonfarm households (Gardner, 1992; Mishra, El-Osta, Morehart, Johnson, & Hopkins, 2002; Ahearn, El-Osta, & Dewbre, 2006). Many studies conducted on US farms, however, have found that payments to farmers have decreased off-farm labor participation (El-Osta & Ahearn, 1996; Mishra & Goodwin, 1997; Ahearn et al., 2006). For example, Ahearn et al. (2006) found that payments from the Conservation Reserve Program, a program that is similar to the *Grain for Green* program, decreased the likelihood of a farm operator working off the farm. These findings suggest that the substitution effect, which would increase off-farm labor allocation, is outweighed by the income effect, which would decrease the number of hours allocated to off-farm labor. Other studies suggest that government payments may decrease off-farm employment because the payments could serve as a means to cope with risk (Key, Roberts, & O'Donoghue, 2006).

---

[2]The study by Groom et al. (2006) uses a household survey implemented in 2004 and collected 1999 pre-program data on a recall basis.

While the findings from the US suggest a hypothesis that *Grain for Green* would lead to decreased off-farm participation, the impact of conservation payments in a rural, developing economy may not follow the same path. Farmers in developing countries have much lower levels of income (and, as such, a higher marginal utility of income) than farmers in the US, so the negative income effect may be small enough that it is outweighed by the positive substitution effect. Moreover, household pre-program participation in off-farm labor markets may be inhibited by low incomes and the absence of liquidity to finance the shift into the off-farm market as well as poorly functioning land and credit markets (Hoff & Stiglitz, 1990; Bardhan & Udry, 1999). Furthermore, farmers in developing economies are more likely to face high transaction costs that prevent them from seeking off-farm employment. Accordingly, if government payments can relax farmers' liquidity constraints, programs may help farmers obtain off-farm jobs.

The literature suggests that this conjecture may apply to rural China. High transaction costs, weak information-sharing and other regulations have been shown to restrict farmers in rural China from starting self-employed enterprises and seeking wage-earning jobs (deBrauw, 2002; Knight & Song, 2005). Case studies suggest that, though formal and informal loans are available, borrowing remains severely constrained, especially for the resource-poor strata of the population (International Fund for Agricultural Development, 2001). Credit constraints have been shown to affect factor allocation in the production decisions of rural China's households (Feder, Lau, Lin, & Luo, 1990). Since land rental markets are frequently incomplete in rural China, most households cannot leave agriculture entirely (Nyberg & Rozelle, 1999). Given these conditions, if the *Grain for Green* program can improve liquidity of farmers, the program may enable them to find off-farm jobs and increase other productive activities. Yet, the extent of this effect may be conditional on the individuals' skills, such as their age and education.

Based on our data, we consistently find that, on average, the *Grain for Green* program has a positive (although only moderate) effect on off-farm labor participation. Households that participate in the program are increasingly allocating their family's labor into the off-farm labor market. The results also indicate that households with less liquidity before participating in the program are more likely to start off-farm jobs, supporting the view that the compensation for setting aside cultivated land may be relaxing liquidity constraints, allowing participants to more readily move into the off-farm sector relative to nonparticipants. However, we also find that the impact is only found among individuals that are younger and have achieved higher levels of education.

In examining the impact of China's *Grain for Green* program, this study is limited in that we examine changes in labor allocation instead of using a more direct measure of welfare such as income. If we had before-after data on income from off-farm labor or the number of days worked off the farm, we could have used those variables as outcomes. Unfortunately, we chose not to collect those variables for 1999. Our first survey of the impact of this program was implemented at a time that was already three years into the program (for some households in the study region). Our pretest revealed that although households remembered quite clearly their participation in the off-farm labor sector, the information that respondents could provide on wage income or the number of days that they worked off the farm was less precise and could potentially suffer from recall bias.[3]

Although the lack of panel data (before and after the program) on wage income is a limitation to answering whether *Grain for Green* is leading to a welfare improvement, changes in labor allocation per se are helpful in understanding the long-run implications of the program. If we find that the program is at least enabling farmers to find off-farm employment, it implies that those farmers are less likely to be stuck in a poverty trap (or remaining in subsistence agriculture) after the program compensation stops.

The literature on off-farm employment in rural China also suggests that participation in the off-farm labor market itself serves as an indicator of welfare improvement (although we realize this is not necessarily true in all countries). Empirical studies have consistently shown that improved access to off-farm jobs has created a large share of the increase in rural incomes in the economic reform era (e.g., Parish, Zhe, & Li, 1995; Zhang, Huang, & Rozelle, 2002; de Janvry, Sadoulet, & Zhu, 2005).[4] Rising off-farm employment also has been linked to reduced poverty (Du, Park, & Wang, 2005). Fan, Zhang, and Zhang (2004) show that the shift into off-farm employment not only results in reduced poverty, but also is a driver of structural change in the economy. In sum, based on the recent emergence and studies of the off-farm labor market in rural China, we believe that increasing off-farm labor itself leads to welfare improvement.

9.2 The *Grain for Green* Program

Starting in 1999 as a pilot program, the *Grain for Green* program was implemented by China's government as a cropland set-aside program to increase forest cover and prevent soil erosion on cultivated slopes.[5] By 2010, the State Forestry Administration (SFA) plans to convert 15 million ha of crop land (approximately 10% of all of China's cultivated area) (State Forestry Administration, 2003).[6] Since the main

---

[3]However, this limitation in fact may strengthen our finding. In a sense, we are doing a stricter test of how many new individuals are working off the farm due to the program. If we indeed find that there are additional new individuals in the off-farm labor market, we conjecture that we would find an increase in the duration as well.

[4]The rural nonfarm sectors consist essentially of Township and Village Enterprises (TVEs) and the rural private economy.

[5]For an excellent overview of *Grain for Green*, see Xu et al. (2004).

[6]Due to recent controversies over fiscal pressures, hikes in grain prices, and delivery of program compensation, the government scaled back expansion of the program in 2005 (Xu et al., 2006). However, in 2007, the government announced that it will extend the program until 2021 with the same goal of converting 15 million ha in total (Guowuyuan, 2007).

objective of China's program is to restore the nation's forests and grasslands to prevent soil erosion, program designers have set slope as one of the main criteria by which plots are selected.

Most close observers believe that *Grain for Green* has been "quasi voluntary." The SFA and provincial and sub-provincial forestry bureaus are primarily responsible for targeting general areas of land for enrollment in the program as well as in setting and distributing enrollment quotas to local governments (Zuo, 2002). In practice, the central and the local governments bargain over the land conversion quota (Xu, Tao, Xu, & Bennett, 2006). Since compensation in most cases exceeds foregone income of cultivation and the funds go through the hands of local implementing agencies and local finance bureaus, local officials frequently overreach the land retirement quota set by the central government as a way to help them bargain for a higher budget. In field interviews, some households reported being "strongly encouraged" to participate. In fact, Xu et al. (2006) found that nearly half of the participating households in the sample believed that they did not have the autonomy to choose whether or not to participate, and only 30% had the autonomy to choose which plots to retire. We exploit the quasi-voluntary nature of program participation in our identification strategy.

According to the program's rules, each participating farmer receives three types of compensation: in-kind grain, cash and free seedlings. In-kind grain and cash are given out annually after a farmer's program plot passes an inspection; seedlings are provided only in the first year. The program is designed so that there are only two levels of compensation nationwide, which reflect inherent differences in regional average yields. The compensation level is 1,500 kg of grain per hectare per year in the Yellow River basin and 2,250 kg/ha/year in the Yangtze River basin. In cash-equivalent terms, the sum of the three types of compensation given to farmers in the upper and middle reaches of the Yellow River basin amounts to 3,150 yuan/ha during the first year of conversion and 2,400 yuan/year/ha in following years.[7] For the upper reaches of the Yangtze River, the program pays farmers 4,200 yuan/ha in the first year and 3,450 yuan/year/ha thereafter. The level of compensation is not trivial relative to the earnings of the typical participating household in the study region. For example, if an average household in Sichuan Province (Yangtze River basin) received full compensation, it would receive 340 yuan per capita, an amount equal to 24% of average per capita income in 1999 (Uchida et al., 2007).[8]

---

[7]The annual average official exchange rate in 2001 was 8.28 Chinese yuan to one U.S. dollar. The purchasing-power parity conversion factor in 2001 was 1.9 yuan to the dollar (World Bank, 2003).

[8]According to Xu et al. (2006), actual compensation received by farm households in our sample fell short of the compensation standards set in the program guidelines for a fraction of the households. Among the households that did not voluntarily participate, they received, on average, only 46% of their promised compensation in 2002, compared with 62% for those who participated voluntarily. There are two plausible reasons for this snortfall in receiving payment. First, based on informal interviews during our field work we found some farmers whose payments were lagging due to logistical reasons. Program expansion had been so fast that local government agencies responsible for program supervision did not have sufficient manpower to check whether the converted land satisfied government-stipulated requirements (such as tree types and survival rates). Second, some

The program can potentially affect household wealth, both directly and indirectly. *Grain for Green* directly affects household incomes through the grain and cash compensation, which can be used for other productive activities and for consumption. Previous studies of the *Grain for Green* program have found that the compensation rate typically is larger than the value of the crop yielded by the retired plots (Uchida, Xu, & Rozelle, 2005; Xu et al., 2006). The conservation set-aside program also can indirectly induce structural change in household wealth by reducing the demand for labor for cultivating crops. How the freed-up labor time gets reallocated may critically depend on the other resources possessed by the household, the household's stock of human capital and the conditions of land, labor and credit markets.

9.3 Data

We use a panel data set from household surveys that we designed and implemented in 2003 and 2005. The surveys were commissioned by China's SFA to evaluate the *Grain for Grain* program. This data set is believed to be the only existing panel data set that includes both participating and nonparticipating households. The descriptive statistics for the key variables discussed here are shown in Table 9.1.

The 2003 household survey used a stratified sampling strategy designed to collect data on a random sample of 359 households in the program area. From the three provinces (Sichuan, Shaanxi and Gansu) that had been participating in *Grain for Green* since 2000, two counties, three townships per county, two participating villages per township and ten households per village were randomly selected.[9] The data include information on at least one program-participating household for each village. Two of the 36 villages had only participating households. The survey in 2003 collected information on 2002 and 1999, and the survey in 2005 collected information on 2004. Of the 359 households surveyed in 2003, we were able to track 270 of them in 2005, 230 of which were participating households. Of the 230 households, 27 entered the program in either 2003 or 2004. The attrition rate (from the survey) was 24% for households participating in the program and 32% for nonparticipating households. However, the households not included in the 2005 survey were not systematically different from households that were included in both surveys. We dropped all of the households that dropped out of the sample in 2005 from the analysis.

Among the program participants, the intensity of their participation, in terms of the number of years they have been in the program and the share of the household's

case studies have found that local governments retained some compensation to make up for expenditure shortfalls and tax arrears. In other cases government kept some of the funds to compensate themselves for expenditures on plant seedlings and other costs.

[9]Based on our sampling strategy, the sample is representative of households in participating counties in the three provinces. For example, when we compare the provincial means of household size and area of cultivated land in 1999, we find them comparable to the statistics published by National Bureau of Statistics of China (1999).

Table 9.1 Descriptive statistics of participating and nonparticipating households

	Participants (as of 2004)	Nonparticipants
Samples in panel data		
No. of households in sample – 1999	0	270
No. of households in sample – 2002	201	69
No. of households in sample – 2004	230	40
No. of individuals in sample – 1999	0	1,010
No. of individuals in sample – 2002	768	242
No. of individuals in sample – 2004	935	155
Program Characteristics – 2004		
Number of years in program (years)	4.5	n.a.
Program area (mu)	9.3	n.a.
Ratio of program area to total land holdings (%)	48.7	n.a.
Household Characteristics – 2002		
Schooling of household head (years)	4.8	4.7
Age of household head (years)	47	48
Number of household members over age 15 (persons)	3.8	3.6
Average age of household members over age 15 (years)	39	41
Average educational attainment of household members over age 15 (years)	4.7	4.4
Number of children (younger than age 15)	1.1	0.9
Total land holdings (mu)	13.7	10.0
Average slope weighted by land area (1 = less than 15 degrees; 2 = 15–25 degrees; 3 = more than 25 degrees)	2.0	1.6
Asset holdings per Capita (1999)		
Livestock assets (yuan)	88	113
Consumer durables (yuan)	461	481
Fixed productive assets (yuan)	231	147
Loans, productive (yuan)	35	25
Loans, consumption (yuan)	459	192
Bank savings (yuan)	42	14
Total asset value (yuan)	1,338	972

Note: Zero values were included when calculating the means for asset holdings per capita.

cultivated area, varied widely across the sample. A third of the households in the sample started to participate during the initial year of the program, but others started later (Fig. 9.1). The share of land that each household retired from cultivation also varied among participating households, ranging from less than 5% of total cultivated land holdings to 100%. We use these two variables as measures of the intensity of program participation and as tools for identifying the effects of the program.

By combining data from the 2003 and 2005 surveys we are able to produce information on labor allocation for both before (in 1999) and after (in 2002 and 2004) the

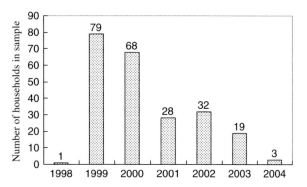

Fig. 9.1 Number of Participating Households by Starting Year. (Note: $n = 230$. Households Started Retiring Cultivated Land at the End of the Harvest Season, So Those Who Said That They Participated in 1999 Actually Retired the Land for 2000.)

implementation of the *Grain for Green* program. Enumerators collected information on each household's production activities on a plot-by-plot basis. The survey also collected detailed information on each household's asset holdings, its demographic make-up and other income-earning activities involving both on-farm and off-farm activities.

The study relies on information for 1999 that was collected in 2003, and we acknowledge the potential for problems inherent in recall data, especially regarding the pre-program period. We addressed concerns about recall bias through the design of the survey. We also carefully trained and monitored the enumerators to ensure that respondents produced their best recollections of past amounts and activities. We also try to deal with the recall bias by reestimating all of the analyses using a sample of individuals from only 67 households—the 27 households that switched from non-participant to participant status between the two surveys (2002 and 2004) and the 40 nonparticipating households. With this subsample, we compare the changes in off-farm labor between 2002 and 2004 to avoid having to rely on the recall data for 1999. We believe since the results from the analysis using the subsample of households are largely consistent with the results from the analysis using the full sample, the recall bias is limited. Full details are available in the Appendix.

9.3.1 Off-Farm Labor Allocation

By 2004, a large share of participating household members had reallocated their time to off-farm work (Fig. 9.2).[10] In the 2005 survey enumerators asked each respondent what the participating household did with the time that was freed up after implementation of the program. The largest share of respondents replied that they had reallocated the time of household members to off-farm work (32%).

---

[10]In this study off-farm labor includes any labor that is not on a farm. We define an individual to have an off-farm occupation if the person engages in wage-earning activities in an off-farm firm or in nonfarm self employment for at least seven days in a given year.

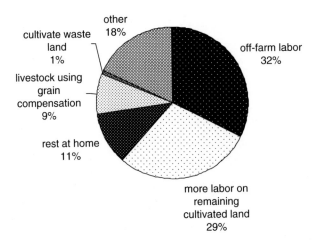

Fig. 9.2 Time Reallocation Choices After Participating in the *Grain for Green* Program. (Note: *n* = 325 Responses from 266 Program Participants (Multiple Choices). Responses to the Question "What Do You Do with the Freed-Up On-Farm Labor Time after Participating in the *Grain for Green* Program?")

The second most frequent response was that households had allocated more labor to their remaining cultivated land (29%). In addition, respondents stated that they had invested this freed labor time in leisure time (or time spent at home—11%) and (in conjunction with the grain compensation) to increase the scope of their livestock enterprises (9%).

Descriptive statistics from the household data showed that off-farm labor allocation was increasing for both participating and nonparticipating households. Between 1999 and 2004, individuals with off-farm jobs increased 13% for participating households and 8% for nonparticipating households. Because off-farm employment is changing for both types of households, it is clear that in order to evaluate convincingly the impact of the program on off-farm labor, we need to control for the time effect and thus cannot simply compare post-program levels of off-farm work between the two groups.

Among the individuals that had off-farm employment in 2002, we find that 42% had jobs that were not local (implying that they were part of the migrant labor force and living and working away from home). Forty percent of individuals with off-farm employment had local wage-earning jobs and 18% were self-employed. We did not have a variable that distinguished between types of off-farm work in the 2005 survey, and thus we relied on the binary variable that indicated whether an individual member had an off-farm job or not. For 2002 and 2004, however, we do have information regarding the intensity of off-farm work (Table 9.2). We find that between these 2 years the average hours worked per day and the number of days per year increased for participants but not for nonparticipants. Earnings from off-farm work and remittances increased for both groups but the differences between the two groups in a particular year are not statistically significant.

Table 9.2 Descriptive statistics of labor allocation for participants and nonparticipants

	Participants (as of 2004)			Nonparticipants (as of 2004)		
	1999	2004	% change	1999	2004	% change
Percent of individuals with off-farm work (%)	23.9	32.4	+9**	28.2	30.8	+2.6**
Percent of individuals with farm work (%)	69.1	67.4	–2**	69.4	76.7	+7.3**
Household members with off-farm work (persons)	0.72	1.24	+72	0.93	1.15	+24
Household members working on-farm (persons)	2.59	2.59	0	2.53	2.90	+15
	2002	2004	% change	2002	2004	% change
If the individual has off-farm work:						
Hours per day	9.2	9.6*	+4	9.4	9.0*	–4
Days per year	171	188	+10	196	164	–26
Months per year	6.7*	7.0	+10	7.6*	6.3	–7
Annual earnings (yuan)	3,313*	4,305	+30	4,339*	5,736	+32
Annual remittances (yuan)	1,936*	2,362	+22	2,812*	3,180	+13

* indicates that the average for participants and nonparticipants for the given year are statistically significantly different.
** indicates that the numbers are percentage points.

While there was a detectable increase in off-farm employment participation for both program participants and nonparticipants, the same cannot be said for on-farm work. Nonparticipants who engaged in some farming activities increased their work on the farm by six percentage points; at the same time participants decreased on-farm work by four percentage points. The reason why on-farm labor did not decrease as much as the increase in off-farm activities may be because individuals frequently work in part-time off-farm jobs. They often return to farm work periodically.

9.4 Identification Strategy

To identify the variations across households in off-farm labor market participation that are due to factors other than the *Grain for Green* program, we exploit the quasi voluntary nature of the program. Many households did not have either the choice of whether or not to participate in the program or the choice about which plot to enroll into the program. Because of this, there is less potential for self-selection. In addition, the program officers that were in charge of selecting who got to participate and which plots were able to be enrolled based their decisions on slope and other characteristics of each household's land holdings (which are both observable.)

Indeed, the weighted average of slopes was higher for participants compared to non-participants (Table 9.1). Accordingly, we first utilize the data from both participating and nonparticipating households in difference-in-differences (DID) estimators. We then employ matching methods where we match participants and nonparticipants by slope and other characteristics of each family's land resources (i.e., those factors that at least in part determined program placement.).

DID compares outcomes from a policy change on two groups—those affected by the policy change (program participants) versus those who are not (nonparticipants, Meyer, 1995). The idea is to correct the simple difference between an outcome before and after a policy change for the treatment group by comparing the before-after change of treated units with the before-after change of control units. By doing so, any *common* trends in the outcomes of both the control and treated units are differenced out (Smith, 2004). The estimator also eliminates recall bias inherent in a retrospective survey as far as the bias is the same for the two groups.

DID differences out all the time-invariant variables—both observed and unobserved—such as the households' total time endowment and other time-invariant household characteristics that determine consumption and production. Next, DID zeroes out any time-variant variables which change in parallel between the two groups. For example, to the extent that input and output prices as well as transaction costs for participation in off-farm labor market move in parallel for the two groups, these effects are captured.

The conventional DID estimator, however, assumes that in the absence of the program, average outcomes for participants and nonparticipants follow parallel trends over time. This assumption may be implausible if unobservable pre-program characteristics are thought to be associated with the dynamics of the outcome variable and the characteristics are different for participating and nonparticipating groups. We therefore report the DID estimates from models that include an interaction term between a dummy variable for the Yangtze River Basin and a year dummy variable for 2004 to capture the differences in compensation rates: a higher level for samples in Sichuan Province (Yangtze River Basin,) and a lower level for households in Gansu and Shaanxi Provinces (Yellow River Basin). This term also captures other systematic differences in changes over time between the two basins. We also include the off-farm job status in 1999 and other pre-program control variables in order to increase the probability that the parallel assumption holds. Following the literature, we include household characteristics that are considered to be important in determining off-farm labor participation, namely, household size, the household's number of children under age 15 and liquidity in 1999.

Given the preceding considerations, we estimate the empirical model as

$$L_{it}^o = \beta_o + \beta_1 T_t + \beta_2 D_t + \beta_3 (T_t * D_i) + \beta_4 (T_t * R_i) + \beta_5 L_{i,t=o}^o + \beta_6 R_i + \gamma X_i + \varepsilon_{it} \quad (9.1)$$

where L_{it}^o is off-farm labor status, D_t is a dummy variable taking the value 1 if individual i is in the treatment group and 0 if it is in the control group, T_t is a dummy taking the value 1 in 2004 and 0 in 1999. The coefficient β_3 is the parameter of interest, the DID estimator. In addition, to better control for different trends over time, we control for off-farm labor participation in base year ($L^o_{it=o}$), a regional dummy which takes the value 1 if the household i is located in the Yangtze River Basin (R_i) and other household characteristics (X_i). We estimate the model using individual data with a probit estimator and report the marginal effects. Since errors may be correlated within households, we allow the standard errors to be clustered at the household level.

We also extend the DID framework to test whether the intensity of participation in the program influences the program effect by replacing the treatment variable D_i with a measure of intensity. If devoting more land to the program led to an increase in available labor time or an increase in liquidity that households could use to find off-farm jobs, the difference in off-farm employment might be positively related to the area of land retired by each household. As measures of program intensity, we examine (a) the ratio of the program area to each household's total land holding; (b) the number of years in the program; and (c) an interaction term between the ratio of retired to total land area and the duration of the program.

In addition to the program's effect on shifts in off-farm employment, we also estimate the effect on changes in each individual's on-farm status. Farmers with improved liquidity may either intensify labor input per hectare cultivated and/or shift labor from on-farm to off-farm employment. Based on our descriptive statistics, some participating farmers also increased their livestock production after the program started, utilizing the in-kind grain compensation for feed. Others switched from conventional grain to higher-value horticultural commodities. Whether or not the former effect offsets the shift towards off-farm labor is an empirical question.

9.4.1 Sensitivity Analysis

There are reasons, however, why we cannot be too confident about the DID approach. Some households did have the autonomy over program participation, in which case we may have a selection problem. Moreover, descriptive statistics suggest that it might be the case that the control group is different in some aspects from the treatment group. These differences may reduce our confidence in the DID estimates.

One strategy to deal with this problem is to match treatment households to control ones with similar characteristics and look at the differences in behavior over time. In the literature, this approach is called difference-in-difference matching (DID matching). The matching method is one way to examine the impact of a treatment on an outcome when selection takes place on observable characteristics (Rosenbaum & Rubin, 1984). Measuring the effect of the program on off-farm labor without bias using DID matching requires that we include all characteristics which (a) determine which households are likely to participate in the program; and (b) also affect changes in labor allocation. If these assumptions hold, we can say that given the observable covariates, changes in the off-farm labor allocation decisions of nonparticipating households are what the level of off-farm labor work of the participating households would have been had they not participated in *Grain for Green*. By combining matching with the DID approach, we can difference away time-invariant unobservable variables as well as time-variant factors that have a parallel trend (Smith & Todd, 2005). We match the two groups of households based on (a) slope and other characteristic of household's land resources that officials used (implicitly and explicitly) for program placement; and (b) factors that affect differences in access to off- and on-farm job opportunities, such as: education, age, household size, number of children, liquidity and crop income in 1999, and the distance from the village to the township seat.

To implement the estimator, we use two covariate matching estimators: (a) a nearest-neighbor covariate matching estimator with an inverse variance weighting matrix (to account for differences in the scale of the covariates); and (b) a nearest-neighbors covariate matching estimator with Mahalanobis weighting (a distance measure that accounts for scale—Abadie & Imbens, 2006). Matching is done between an observation and its nearest neighbor with replacement. We resolve the mean-variance tradeoff in the match quality by using two nearest neighbors; the counterfactual outcome is the average among these two. To deal with the bias in finite samples when the matching is not exact, we use the post-matching bias-correction procedure of Abadie and Imbens (2006) that asymptotically removes the conditional bias term in finite samples. We also compare the results with propensity score matching using kernel matching. Since the matching estimator works well only when the treatment variable is binary, we only estimate the effect of the discrete choice of program participation.

9.4.2 Strategy to Estimate How Liquidity Affects the Program's Impact on Off-Farm Labor

Two of our variables that can be used as measures of liquidity also depict different trends between the participating and nonparticipating groups. Since we are specifically interested in whether the program's effect on labor allocation differs for households with different levels of liquidity, we turn now to the strategy for testing this proposition.[11] Ideally, if we could directly classify households into those that were liquidity-constrained and those that were not (e.g., Zeldes, 1989; Carter & Olinto, 2003), we could estimate the program's impact for each group and test whether there were statistically detectable differences between the two groups. Unfortunately, we do not have sufficient information on credit and loan application history from the surveys to do this.

---

[11] The reliability of the DID estimator lies in the identification assumption that there are no omitted time-varying effects that are correlated with program participation. For example, the identification assumption might be violated if other local governmental programs existed that both affected labor allocation and were correlated with participation in *Grain for Green*. Unfortunately, we did not have information to control for other governmental programs and thus the reader needs to interpret all results with this caveat in mind.

Consequently, we take two alternative approaches. We first calculate the pre-program value of each household's liquid assets. In the paper, our definition of liquid assets includes the value of livestock assets, fixed productive assets and consumable durable goods, plus loans and deposits. We then divide the sample households into quartiles based on the value of their total liquid assets: $Q_j, j = [1,2,3,4]$, where $Q_j = 1$ if the household belongs to quartile j and 0 otherwise, and $j = 1$ is the group of households with the lowest asset value. We then test whether the program effects differ among the quartiles using the DID framework. Heterogeneity in treatment effects can be studied by including interactions between Q_j and the treatment dummy variable. Thus, we estimate the following equation:

$$L_{it}^o = \beta_o + \beta_1 T_t + \beta_2 D_t + \sum_{j=1}^{4} \beta_{i,j}(Q_j * T_t * D_i) + \beta_4(T_t * R_i) + \beta_5 L_{i,t=o}^o + \beta_6 R_i + \gamma X_i + \varepsilon_{it}$$

(9.2)

If a household's liquidity constraint is indeed being relaxed by participation in *Grain for Green*, there will be a positive impact by the program on participation in the off-farm labor market (or on earnings from agriculture). In the empirical model we anticipate that households that had a lower level of liquidity before *Grain for Green* (those households belonging to the lower two quartiles) will see a greater relaxing of their liquidity constraint when they receive their compensation than households that owned sets of liquid assets with higher values (or those from the top two quartiles).

As a second alternative approach, we utilize a rule developed by Zeldes (1989) to split the households into liquidity-constrained and -unconstrained groups and compare the DID estimates between the two groups.

9.5 Effect of the *Grain for Green* Program on Off-Farm Labor

Point estimates from the DID model reveal that the *Grain for Green* program increased off-farm labor participation and decreased on-farm labor participation (Table 9.3, Column 1). Off-farm labor participation increased for both participants and nonparticipants, but it increased more for participating households. Participating in the program increases the likelihood of an individual person working off-farm by 15 percentage points, an estimate that is statistically significant at the 5% level. The sign of the coefficient estimated using the basic DID model suggests that *Grain for Green* is promoting structural change. Based on this result, for those that expect that *Grain for Green* will help to promote off-farm employment, the results of the basic regression models are somewhat encouraging.

The results also reveal that when program participation is done with greater intensity, it increases off-farm labor participation (Table 9.3, Columns 2–4). Specifically, a higher ratio of retired land to total holdings and its interaction term with the number of years in program lead to a greater propensity of obtaining an off-farm job. The results suggest that an individual from a household that retires all of its

Table 9.3 Impact of *Grain for Green* on off-farm labor decisions, 1999 and 2004

	(1)	(2)	(3)	(4)
	Dependent variable: 1 = Off-farm work, 0 = No off-farm work			
Treatment × year 2004	0.149 (2.02)**			
Ratio of program area to total land holding × year 2004		0.159 (3.51)***		
Number of years in program × year 2004			0.020 (1.72)*	
Ratio of program area to total land × number of years in program × year 2004				0.029 (3.68)***
Treatment	-0.038 (1.38)			
Ratio of program area to total land		-0.083 (3.01)***		
Number of years in program			-0.007 (1.48)	
Ratio of program area to total land × number of years in program				-0.015 (3.14)***
Year 2004 dummy	0.068 (0.97)	0.119 (3.26)***	0.117 (2.14)**	0.124 (3.64)***
Year 2004* Yangtze basin	0.008 (0.17)	0.026 (0.53)	0.002 (0.04)	0.026 (0.52)
Yangtze basin dummy	0.047 (2.23)**	0.033 (1.61)	0.047 (2.29)**	0.033 (1.64)
Household size	0.011 (1.49)	0.010 (1.22)	0.010 (1.37)	0.010 (1.19)
Number of children age<15	-0.006 (0.48)	-0.003 (0.23)	-0.004 (0.37)	-0.003 (0.23)
Total land holding	-0.001 (0.88)	-0.001 (0.88)	-0.001 (0.86)	-0.001 (0.88)
Age in 2002	-0.006 (8.57)***	-0.006 (8.67)***	-0.006 (8.63)***	-0.006 (8.65)***
Education in 2002	0.010 (3.55)***	0.010 (3.45)***	0.010 (3.46)***	0.010 (3.45)***
Total household liquidity in 1999	0.000 (1.67)*	0.000 (1.70)*	0.000 (1.51)	0.000 (1.69)*
Individual had off-farm work in 1999	0.731 (22.67)***	0.728 (23.18)***	0.728 (22.83)***	0.728 (23.16)***
Pseudo R-squared	0.44	0.44	0.44	0.44
Observations	1953	1953	1953	1953

* significant at 10%;** significant at 5%;*** significant at 1%

Note: The reported coefficients are marginal effects of probit models. Robust z-statistics in parentheses are calculated based on the clustered standard error at the household level.

cultivated land will increase the likelihood of an adult member working off the farm by 16 percentage points (Column 2). Duration in the program, by itself, is only weakly associated with greater off-farm labor participation (Column 3), but when it is jointly considered with program area, the longer a household has been in the program with a higher proportion of its cultivated land, the greater its increases in off-farm employment (Column 4). Interestingly, we find that program intensity matters for changes in off-farm labor participation but not for changes in on-farm labor (Table 9.4, Columns 1–4). This result may be due to the discrete measure of on-farm work, which does not capture changes in time spent on-farm.

Overall, the findings from the smaller subset are consistent with those from the full sample. Full results are available in the Appendix.

9.5.1 DID Matching Results

The estimates of the program's effect on off-farm employment obtained using different variations of the DID matching methods are consistent with the conventional DID estimates: *Grain for Green* has increased the likelihood of an adult member working off the farm (Table 9.5, Column 1). The point estimates ranged from 15 percentage points to 20 percentage points. For example, based on covariate matching using inverse distance, program participation increased the likelihood of an individual working off-farm by 19 percentage points, an estimate that is statistically significant at the 5% level (Row 1). However, the DID matching estimates for the change in on-farm job status were different from the estimates produced by a DID estimator. Whereas the DID estimates were statistically insignificant for changes in on-farm job status, the DID matching results were consistently negative and mostly statistically significant, ranging from 11 to 16 percentage points (Table 9.5, Column 2). The DID matching results suggest that program participation decreased the likelihood of an adult member working on the farm.

9.6 Heterogeneous Program Effect on Off-Farm Labor

In the previous section, we found that the *Grain for Green* program led to an increase in off-farm labor participation. The DID estimates, however, do not allow us to understand *how* the program affects off-farm labor or which types of farmers are able to shift into the off-farm sector. In fact, we are interested in understanding how these changes occur. In particular, when households make off-farm labor-participation decisions we want to understand the role of two factors: physical capital and human capital. In this section we test whether *Grain for Green* has heterogeneous effects on off-farm labor that depend on the availability of physical and human capital to the households before the program.

Table 9.4 Impact of *Grain for Green* on farm labor decisions, 1999 and 2004

	Dependent variable: 1 = On-farm work, 0 = No on-farm work			
	(1)	(2)	(3)	(4)
Treatment × year 2004	-0.133 (1.66)*			
Ratio of program area to total land holding × year 2004		0.023 (0.38)		
Number of years in program × year 2004			-0.011 (0.78)	
Ratio of program area to total land × number of years in program × year 2004				0.003 (0.26)
Treatment	0.017 (0.68)			
Ratio of program area to total land		-0.025 (1.01)		
Number of years in program			-0.000 (0.07)	
Ratio of program area to total land × number of years in program				-0.005 (1.08)
Year 2004 dummy	0.123 (1.59)	-0.006 (0.13)	0.050 (0.80)	-0.001 (0.03)
Year 2004* Yangtze Basin	-0.085 (1.24)	-0.058 (0.86)	-0.073 (1.10)	-0.060 (0.90)
Yangtze Basin dummy	0.041 (2.17)**	0.033 (1.70)*	0.039 (2.06)**	0.033 (1.73)*
Household size	-0.018 (2.38)**	-0.018 (2.33)**	-0.018 (2.30)**	-0.018 (2.32)**
Number of children age<15	0.018 (1.14)	0.014 (0.96)	0.016 (1.02)	0.014 (0.93)
Total land holding	-0.000 (0.06)	-0.000 (0.14)	-0.000 (0.07)	-0.000 (0.06)
Age in 2002	-0.001 (0.94)	-0.001 (0.90)	-0.001 (0.92)	-0.001 (0.91)
Education in 2002	-0.005 (1.49)	-0.005 (1.44)	-0.005 (1.50)	-0.005 (1.45)
Total household liquidity in 1999	-0.000 (1.44)	-0.000 (1.46)	-0.000 (1.37)	-0.000 (1.45)
Individual worked on farm in 1999	0.716 (27.51)***	0.714 (27.80)***	0.714 (27.58)***	0.713 (27.74)***
Pseudo R-squared	0.44	0.44	0.44	0.44
Observations	1955	1955	1955	1955

* significant at 10%;** significant at 5%;*** significant at 1%
Note: The reported coefficients are marginal effects of probit models. Robust z-statistics in parentheses are calculated based on the clustered standard error at the household level.

Table 9.5 Program impact on off-farm and farm job status change between 1999 and 2004 using covariate matching method

	Outcome: Off-farm status in 2004 – Off-farm status in 1999	Outcome: On-farm status in 2004 – On-farm status in 1999
Covariate matching (inverse variance)	0.192** (2.260)	–0.159* (1.85)
Covariate matching (Mahalanobis)	0.197** (2.74)	–0.150* (2.33)
Propensity score matching (kernel, common support)	0.147** (3.24)	–0.112* (1.88)
N treated	818	818
N available controls	138	138

Note: For covariate matching, we report the absolute value of Abadie and Imbens' bias-adjusted z-statistics in parentheses. For the propensity score matching estimator we report the z-statistics after bootstrapping the standard error 999 times. Calipers restrict matches to units within 0.5 standard deviations from each covariate. Propensity score matching was performed based on kernel matching (with Epanechnikov kernel.) A common support was imposed, which resulted in 22 treated observations off the common support. Covariate matching and estimation of the propensity score was performed using the following variables: age, education, household size, number of children younger than 15, total land holding, liquidity asset level in 1999, average slope weighted by land area, average distance from plots to a public road weighted by land area, average soil quality weighted by land area, average distance from plots to a waterway weighted by land area and crop income in 1999.
** = significant at 5% level; * = significant at 10% level

9.6.1 Liquidity Constraint

We find that the effect of the program on off-farm labor is clearly larger for households that had less liquid assets prior to the program (Table 9.6, Column 1). For participants belonging to the lowest quartile, the propensity to work off farm increased by an average of 23 percentage points. The magnitude of the coefficients gets steadily smaller as asset quartiles rises and was not statistically significantly different from zero for the richest quartile. In contrast, estimates for on-farm work suggest that individuals in the lowest quartiles moved away from on-farm work (Column 2). The magnitude of the coefficient is smaller for households in higher quartiles. We also found consistent results when we split the households using Zeldes' rule into liquidity-constrained and -unconstrained groups and compared the DID estimates. Full results are available in the Appendix.

In sum, the findings reveal that the more liquidity-constrained a household is prior to the program, the more positive the impact of the *Grain for Green* program is on its off-farm employment participation. One way of interpreting this result is that participation in the program relaxes a household's liquidity constraint and that the household uses its new resources to aid it in its decision to participate in off-farm work. Thus, the more constrained is the household, the larger is the program's impact on off-farm work.

Table 9.6 Program impact on off-farm and farm jobs, treatment indicator interacted with quartile dummies of asset holdings, education and age, 1999 and 2004

	(1) Off-farm	(2) On-farm	(3) Off-farm	(4) On-farm	(5) Off-farm	(6) On-farm
Poorest in asset value in 1999 (dummy) × treatment × year 2004	0.227 (2.80)***	−0.165 (1.97)**				
Second poorest in asset value in 1999 (dummy) × treatment × year 2004	0.214 (2.70)***	−0.081 (1.02)				
Second richest in asset value in 1999 (dummy) × treatment × year 2004	0.128 (1.67)*	−0.116 (1.41)				
Richest in asset value in 1999 (dummy) × treatment × year 2004	0.115 (1.52)	−0.150 (1.82)*				
Youngest age group (dummy) × treatment × year 2004			0.360 (4.26)***	−0.099 (1.23)		
Second youngest age group (dummy) × treatment × year 2004			0.182 (2.35)**	−0.102 (1.25)		
Second oldest age group (dummy) × treatment × year 2004			0.143 (1.88)*	−0.083 (1.00)		
Oldest age group (dummy) × treatment × year 2004			−0.138 (2.40)**	−0.275 (3.10)***		
Least education (dummy) × treatment × year 2004					0.020 (0.28)	−0.179 (2.17)**

Table 9.6 (continued)

	(1) Off-farm	(2) On-farm	(3) Off-farm	(4) On-farm	(5) Off-farm	(6) On-farm
Second least education (dummy) × treatment × year 2004					0.139 (1.88)*	−0.093 (1.18)
Second highest education (dummy) × treatment × year 2004					0.226 (2.91)***	−0.121 (1.52)
Highest education (dummy) × treatment × year 2004					0.205 (2.41)**	−0.055 (0.65)
Observations	1926	1928	1926	1928	1930	1932

Note: In addition to the treatment variables indicated above, all models also control for year 2004 dummy, Yangtze Basin dummy, household size, number of children age<15, total land holding, total household liquidity asset in 1999, education in 2002, age in 2002, and whether or not individual had off-farm work in 1999. We report the marginal effects of probit models. The standard errors are clustered at the household level. Absolute value of z statistics in parentheses.
* significant at 10%;** significant at 5%;*** significant at 1%.

9.6.2 Human Capital

We also are interested in understanding how human capital can influence the program's effects among households. Age and education are two fundamental indicators of human capital that affect the ability of individuals to find off-farm work. Higher education is expected to result in greater rewards from off-farm labor (Becker, 1993). Education is defined as the number of completed years of schooling and is assumed to capture the skills that individuals may bring to a given job in the off-farm labor market. Previous studies also found that migration is influenced inversely by age; older people are less likely to migrate since they have less time to pay back the investment (Lanzona, 1998). To test whether the program's effect on off-farm labor is influenced by human capital, we divide the sample into quartiles based on an initial level of education and on age cohorts.

The results show that levels of human capital, in terms of both age and education, impact how the program affects off-farm labor (Table 9.6, Columns 3–6). The estimates imply that adult family members who are younger are more likely to shift to the off-farm labor market after the onset of the program than are older ones. For example, for adults in the youngest quartile, the program increased the probability of off-farm labor participation by 36 percentage points; for the oldest quartile, *Grain for Green* decreased off-farm employment by 13 percentage points (Columns 3 and 4). In fact, we believe that this result is plausible considering that the types of off-farm jobs that are most frequently available to farmers are physically demanding (such as construction work) and naturally favor young adults.

Perhaps more importantly, the results show that *Grain for Green* did not have a positive effect on off-farm employment for adults who had only limited education prior to the program (Columns 5 and 6). If the individual was in the lowest quartile for education, participation in the program did not change the likelihood of that person gaining an off-farm job. The results do demonstrate, however, that the likelihood of finding off-farm employment increases as educational attainment increases. This result is consistent with the literature on off-farm labor participation among rural households in China which has found that the level of education attainment affects the ability of households to take advantage of off-farm employment opportunities (e.g., deBrauw, 2002; Zhang et al., 2002; Yang, 2004; de Janvry et al., 2005; Wang, Herzfeld, & Glauben, 2007). Our finding suggests that the program may not be able to induce structural change in income generating activities if participants do not have adequate education for off-farm work. These findings also add yet another piece of empirical evidence suggesting that China will have to expand its investment in education to achieve its goals.

9.7 Conclusion

In our study, we consistently find that, on average, *Grain for Green* has a positive effect on off-farm labor participation. Participating households are increasingly shifting their labor endowment from on-farm work to the off-farm labor market.

This shift occurs not only in absolute terms but is statistically significant when compared to similar shifts in nonparticipating households. In terms of program intensity, we find that program impacts increase as the ratio of a household's retired plots to total land holdings grows. These results, in conjunction with findings from previous case studies on environmental benefits of the program, suggest that *Grain for Green* could potentially be a win-win strategy for achieving both the environmental and poverty alleviation goals.

This finding is in sharp contrast to two prior studies of the *Grain for Green* program that found no effect on off-farm labor participation or on income from off-farm work (Xu et al., 2004; Uchida et al., 2007). Since those studies used household surveys that collected information on labor allocation decisions only for the first 3 years of the program, it may have been too soon for changes to be detected. In this study, we use data collected 5 years after the program began, which may have allowed sufficient time for participants to find off-farm employment in numbers that are statistically detectable.

The results also indicate that households with less liquidity prior to the program are more likely to begin to participate in the off-farm employment market. Our results supports the view that the compensation paid by *Grain for Green* for setting aside cultivated land may be relaxing the liquidity constraints for participating households, allowing participants to more readily move into the off-farm employment sector (relative to nonparticipants).

The positive impact of the program on off-farm labor also is in stark contrast to findings from studies of the impact of government farm payment programs in the US Previous US studies of government payments to farmers, including the Conservation Reserve Program, have typically found that government payments *decrease* household off-farm employment participation. The higher level of income of US farmers compared to what is typical for farmers in the *Grain for Green* program in China probably is the most likely reason why farmers in the US do not choose to work off-farm when offered a government payment (i.e., the wealth effect dominates).

The sensitivity of *Grain for Green* on the level of the household's human capital indicates that there may be more impediments to participating in off-farm labor in rural China than there are in the US. Therefore, in terms of policy impacts for China, if policymakers want to achieve a win-win outcome from *Grain for Green* by meeting both environmental and development goals, they may need to provide additional support to vulnerable populations through job training programs or other means.

Acknowledgements An earlier version of this manuscript has appeared in the American Journal of Agricultural Economics (2009, 91(1): 70–86). The authors are grateful for the assistance in survey design, data collection and entry by Michael Bennett, Yazhen Gong, Zhigang Xu, Fujin Yi, and other members of the Center for Chinese Agricultural Policy, Institute of Geographical Sciences and Natural Resources Research, Chinese Academy of Sciences. We thank Stephen Vosti, James Wilen, Amelia Blanke, Susan Richter, Aslihan Arslan, the Co-editor and three anonymous referees for their comments and suggestions. The authors acknowledge the support from the Ford Foundation, National Science Foundation of China (70024001) and the Agricultural Extension Service of Rhode Island (AES contribution no. 5202) for financial assistance.

9.8 Appendix: Assessing Recall Bias

The study relies on information for 1999 that was collected in 2003, and we acknowledge the potential for problems inherent in recall data, especially regarding the pre-program period. Long-term recall data are potentially inaccurate, although this issue continues to be debated in the literature. Unfortunately, the Chinese government's quick decision to implement *Grain for Green* and lack of transparency in the details of its implementation precluded interviews with potential participants at the program's onset. We also endeavor to deal with the recall bias by reestimating all of the analyses using a sample of individuals from only 67 households—the 27 households that switched from nonparticipant to participant status between the two surveys and the 40 nonparticipating households. With this subset, while the sample is smaller, the data are true panel data and are not subject to errors due to recall. With this subsample, we compare the changes in off-farm labor between 2002 and 2005 to avoid having to rely on the recall data for 1999. If the results from the analysis using the subsample are consistent with the results from the analysis using the full sample, it would suggest that that recall bias is limited.

Overall, the findings from the smaller subset are consistent with those from the full sample (Tables 9.7 and 9.8). The DID estimates for the subset are slightly larger than the estimates for the full sample. This consistency between samples suggests that recall bias in 1999 was limited and/or that the DID approach was able to control for the bias that existed in both groups.

Table 9.7 Impact of *Grain for Green* on Individual Members' On- and Off-Farm Labor Job, Restricting Treated Sample to Participating Households that Changed Status from Nonparticipating to Participating Between 2002 and 2004

	(1) Off-farm	(2) On-farm
Treatment × year 2004	0.325 (2.95)***	−0.037 (0.30)
Treatment	−0.049 (1.31)	−0.027 (0.76)
Year 2004 dummy	0.007 (0.08)	0.128 (1.33)
Year 2004* Yangtze basin	0.178 (1.72)*	−0.120 (0.95)
Yangtze basin dummy	−0.028 (0.72)	−0.004 (0.12)
Household size	0.013 (0.74)	0.022 (1.18)
Number of children age<15 in 2002	0.012 (0.48)	−0.001 (0.03)
Total land holding	−0.000 (0.13)	0.007 (2.71)***
Age in 2002	−0.006 (4.55)***	−0.003 (1.58)
Education in 2002	0.008 (1.21)	−0.010 (2.06)**
Total household liquidity in 1999	0.000 (3.73)***	−0.000 (3.05)***
Individual had off-farm work in 1999	0.741 (10.32)***	
Individual worked on farm in 1999		0.745 (13.73)***
Pseudo R-squared	0.46	0.52
Observations	459	459

* significant at 10%;** significant at 5%;*** significant at 1%
Note: The reported coefficients are marginal effects of probit models. Robust z-statistics in parentheses are calculated based on the clustered standard error at the household level.

Table 9.8 Program Impact on Off-Farm and Farm Jobs, Treatment Indicator Interacted with Quartile Dummies of Asset Holdings, Restricting Treated Sample to Individuals that Changed Status from Nonparticipant to Participant Between 2002 and 2004

	(1) Off-farm	(2) On-farm
Poorest in asset value in 1999 (dummy) × treatment × year 2004	0.398 (2.76)***	–0.061 (0.49)
Second poorest in asset value in 1999 (dummy) × treatment × year 2004	0.460 (2.74)***	0.079 (0.61)
Second richest in asset value in 1999 (dummy) × treatment × year 2004	0.271 (1.64)	0.113 (1.14)
Richest in asset value in 1999 (dummy) × treatment × year 2004	0.058 (0.40)	–0.118 (0.82)
Treatment	–0.055 (0.72)	–0.027 (0.39)
Year 2004 dummy	0.006 (0.08)	0.127 (1.71)*
Year 2004* Yangtze Basin	0.183 (1.68)*	–0.121 (1.17)
Yangtze Basin dummy	–0.022 (0.30)	–0.008 (0.12)
Household size	0.011 (0.49)	0.020 (1.00)
Number of children age<15 in 2002	0.021 (0.72)	–0.010 (0.32)
Education in 2002	0.011 (1.68)*	–0.011 (1.76)*
Age in 2002	–0.006 (3.40)***	–0.003 (2.03)**
Total land holding	0.001 (0.34)	0.008 (2.48)**
Individual had off-farm work in 1999	0.737 (12.05)***	
Individual worked on farm in 1999		0.765 (13.11)***
Pseudo R-squared	0.47	0.53
Observations	453	453

* significant at 10%;** significant at 5%;*** significant at 1%
Note: The reported coefficients are marginal effects of probit models. Robust z-statistics in parentheses are calculated based on the clustered standard error at the household level.

9.8.1 Heterogeneous Program Effect on Off-Farm Labor Using Zeldes' Rule

We found consistent results when we split the households using Zeldes' rule into liquidity-constrained and -unconstrained groups and compared the DID estimates (Zeldes, 1989). The DID estimates for the constrained group was positive and statistically significant both at the household and individual levels. The DID estimates for the unconstrained group were insignificant. The number of participating households that were liquidity-constrained and unconstrained were 170 and 55, respectively, and for non-participating households 32 and 8. At the individual level, the number of participating individuals that were liquidity constrained and unconstrained were 1,316 and 478, respectively, and for non-participating individuals 226 and 72, respectively.

The DID estimates for liquidity-constrained and unconstrained individuals were 0.180 ($z = 2.46$) and –0.049 ($z = 0.31$), respectively. The findings are consistent with the results from the quartile approach.

References

Abadie, A., & Imbens, G. W. (2006). Large sample properties of matching estimators for average treatment effects. *Econometrica, 74*(1), 235–67.

Ahearn, M. C., El-Osta, H., & Dewbre, J. (2006). The impact of coupled and decoupled government subsidies on off-farm labor participation of U.S. farm operators. *American Journal of Agricultural Economics, 88*(2), 393–408.

Alix-Garcia, J., de Janvry, A., & Sadoulet, E. (2003). *Payments for Environmental Services: To whom, where and how much?* (INE/CONAFOR/World Bank, Workshop on Payment for Environmental Service.)

Bardhan, P., & Udry, C. (1999). *Development economics.* Oxford: Oxford University Press.

Becker, G. (1993). *Human capital* (3rd ed.). Chicago: University of Chicago Press.

Bowlus, A., & Sicular, T. (2003). Moving towards markets? Labour allocation in rural China. *Journal of Development Economics, 71*, 561–583.

Carter, M. R., & Olinto, P. (2003). Getting institutions "right" for whom? Credit constraints and the impact of property rights on the quantity and composition of investment. *American Journal of Agricultural Economics, 85*(1), 173–186.

de Janvry, A., Sadoulet, E., & Zhu, N. (2005). *The role of non-farm incomes in reducing rural poverty and inequality in China.* Working paper, Department of Agricultural and Resource Economics, University of California, Berkeley.

deBrauw, A. (2002). *Three essays on migration, education and household development in rural China.* Working paper, University of California, Davis.

Du, Y., Park, A., & Wang, S. (2005). Migration and rural poverty in China. *Journal of Comparative Economics, 33*(4), 688–709.

El-Osta, H., & Ahearn, M. (1996). Estimating the opportunity cost of unpaid farm labor for US farm operators. Technical Bulletin 1848. Washington, DC: US Department of Agriculture, Economic Research Service.

Fan, S., Zhang, L., & Zhang, X. (2004). Reforms, investment and poverty in rural China. *Economic Development and Cultural Change, 52*, 395–421.

Feder, G., Lau, L., Lin, J., & Luo, X. (1990). The relationship between credit and productivity in Chinese agriculture: a microeconomic model of disequilibrium. *American Journal of Agricultural Economics, 72*(4), 1151–1157.

Gardner, B. L. (1992). Changing economic perspectives on the farm problem. *Journal of Economic Literature, 30*(1), 62–101.

Groom, B., Grosjean, P., Kontoleon, A., & Swanson, T. (2006). *Relaxing rural constraints: A 'win-win' policy for poverty and environment in China*? Draft, School of Oriental and African Studies.

Guowuyuan. (2007). Issuance of notification regarding completion of the Sloped Land Conversion Program by the State Department, *Guowuyuan Guanganwanshan Tuigenhuanlinzhengce de Tongzhiguofa*, No. 25.

Hoff, K., & Stiglitz, J. E. (1990). Imperfect information and rural credit markets – puzzles and policy perspectives. *World Bank Economic Review, 4*, 235–250.

Hyde, W., Belcher, B., & Xu, J. (2003). *China's forests: Global lessons from market reforms.* Washington, DC: Resources for the Future.

International Fund for Agricultural Development. (2001). *Rural financial services in China: Thematic study.* International Fund for Agricultural Development.

Key, N., Roberts, M., & O'Donoghue, E. (2006). Risk and farm operator labour supply. *Applied Economics, 38*(5), 573–586.

Knight, J., & Song, L. (2005). *Towards a labour market in China.* Oxford: Oxford University Press.

Lanzona, L. A. (1998). Migration, self-selection and earnings in Philippine rural communities. *Journal of Development Economics, 56*(1), 27–50.

Mayrand, K., & Paquin, M. (2004). *Payments for environmental services: A survey and assessment of current schemes.* Montreal: Unisfera International Centre.

Meyer, B. D. (1995). Natural and quasi-experiments in economics. *Journal of Business and Economic Statistics, 13*(2), 151–161.

Mishra, A., El-Osta, H. S., Morehart, M. J., Johnson, J. D., & Hopkins, J. W. (2002). *Income, wealth, and the economic well-being of farm households* (Agr. Econ. Rep. 812). Washington, DC: US Department of Agriculture, ERS.

Mishra, A., & Goodwin, B. (1997). Farm income variability and supply of off-farm labor. *American Journal of Agricultural Economics, 79*, pp. 880–887.

National Bureau of Statistics of China. (1999). *China statistical yearbook.* Beijing: China Statistics Press.

Nyberg, A., & Rozelle, S. (1999). *Accelerating China's rural transformation.* Washington, DC: The World Bank.

Pagiola, S., Landell-Mills, N., & Bishop. J. (2002). Market-based mechanisms for forest conservation and development. In S. Pagiola, J. Bishop, & N. Landell-Mills (Eds.), *Selling forest environmental services: Market-based mechanisms for conservation and development.* London, UK: Earthscan Publications Ltd.

Pagiola, S., & Platais, G. (2005). Can payments for environmental services help reduce poverty? An exploration of the issues and evidence to date from Latin America. *World Development, 33*(2), 237–253.

Parish, W. L., Zhe, X., & Li, F. (1995). Nonfarm work and marketization of the Chinese countryside. *The China Quarterly, 143*, 697–730.

Rosenbaum, P., & Rubin, D. B. (1984). Reducing bias in observational studies using subclassification on the propensity score. *Journal of the American Statistical Association, 79*, 516–524.

Smith, J. (2004). Evaluating the local economic development policies: Theory and practice. In A. Nolan & G. Wong (Eds.), *Evaluating local economic and employment development.* Paris: OECD Publishing.

Smith, J., & Todd, P. (2005). Does matching overcome Lalonde's critique of nonexperimental estimators? *Journal of Econometrics, 125*, 305–353.

State Forestry Administration. (2003). *Master plan for the sloping land conversion program.* Beijing.

Uchida, E., Xu, J., & Rozelle, S. (2005). Grain for green: Cost-effectiveness and sustainability of China's conservation set-aside program. *Land Economics, 81*, 247–264.

Uchida, E., Xu, J., Xu, Z., & Rozelle, S. (2007). Are the Poor Benefiting from China's Conservation Set-aside Program? *Environmental and Development Economics, 12*, 593–620.

Wang, X., Herzfeld, T., & Glauben, T. (2007). Labor allocation in transition: Evidence from Chinese rural households. *China Economic Review, 18*, 287–308.

World Bank. (2003). *World development indicators.* Washington, DC: The World Bank.

Xu, J., Tao, R., Xu, Z., & Bennett, M. T. (2006). *China's sloping land conversion program: Does expansion equal success?* Unpublished, Peking University.

Xu, Z., Bennett, M. T., Tao, R., & Xu, J. (2004). China's Sloping Land Conversion Programme four years on: Current situation, pending issues. *International Forestry Review, 6*, 317–326.

Yang, D. T. (2004). Education and allocative efficiency: Household income growth during rural reforms in China. *Journal of Development Economics, 74*, 137–162.

Zeldes, S. P. (1989). Consumption and liquidity constraints: An empirical investigation. *The Journal of Political Economy, 97*(2), 305–346.

Zhang, L., Huang, J., & Rozelle, S. (2002). Employment, emerging labor markets, and the role of education in rural China. *China Economic Review, 13*, 313–328.

Zhang, L., Luo, R., Liu, C., & Rozelle, S. (2006). Investing in rural China: Tracking China's commitment to modernization. *The Chinese Economy, 39*(4), 57–84.

Zuo, T. (2002). Implementation of the SLCP. In J. Xu, E. Katsigris, & T. A. White (Eds.), *Implementing the natural forest protection program and the sloping land conversion program: Lessons and policy recommendations.* Beijing: China Forestry Publishing House.

Chapter 10
An Empirical Analysis of the Effects of China's Land Conversion Program on Farmers' Income Growth and Labor Transfer

Shunbo Yao, Yajun Guo, and Xuexi Huo

Abstract In this chapter, we hypothesize that in addition to participation status and household characteristics, the impacts of China's Sloping Land Conversion Program on income growth and labor transfer are determined by the local economic condition, program extent, and political leadership; and the income impacts may vary from sector to sector. To test these propositions, we compiled a dataset of 600 households in three counties of the Loess Plateau region, with observations for times both prior to and after the program initiation (1999 and 2006), both aggregate and categorical incomes, and both participating and non-participating households. Using a difference in differences model and the repeated cross-sectional data, we find that participation status, local economic condition, program extent, and political leadership have indeed made significant impacts on household income and off-farm employment. Moreover, the effects of participation on crop production income, animal husbandry income, and off-farm income vary substantially. These results carry major policy implications in terms of how to improve the effectiveness and impacts of ecological restoration efforts in and outside of China.

Keywords Sloping Land Conversion Program · Income increase · Labor transfer · Repeated cross-sectional data · Northern Shaanxi · Difference in differences model · Economic development · Program extent · Political leadership

10.1 Introduction

The Sloping Land Conversion Program, or SLCP, is a primary national program that has been launched by the Chinese government to mitigate soil erosion, desertification, and other ecological problems in order to achieve more sustainable development. In 1999, the pilot projects of this program were carried out in Shanxi, Gansu, and Sichuan. By the end of 2006, it has subsidized 32.5 million farm households

S. Yao (✉)
College of Economics and Management, Northwest A&F University, Yangling,
Shaanxi 712100, P.R. China
e-mail: yaoshunbo@126.com

R. Yin (ed.), *An Integrated Assessment of China's Ecological Restoration Programs*,
DOI 10.1007/978-90-481-2655-2_10, © Springer Science+Business Media B.V. 2009

in more than 2,200 counties to retire and convert degraded (sloping) and deserti-fied croplands (State Forestry Administration, or SFA, 2007). Its ultimate goal is to convert 14.7 million ha of croplands to forest and grass coverage by the end of this decade, with a total investment of over 220 billion yuan (Yin, Yin, & Li, 2008).[1] The government claims that the program has made a predominantly positive impact on rural households' production and livelihoods as well as on the environment (SFA, 2007). The objective of this chapter is to assess whether or not implementing the SLCP has indeed led to an increase in farmers' income and a transfer of labor into off-farm sectors, and what the key conditions are in determining the program out-come.

Ever since the time when the SLCP was officially announced, its effectiveness and sustainability have been hotly debated. While the government has held a rosy view, scholars have found divergent and even contradicting evidence of the SLCP's impacts. Based on household data collected from Gansu and other provinces and descriptive statistics, Zhi (2004) showed that implementing the SLCP has promoted the transfer of rural labor out of the farming sector and the improvement of farm-ers' income. The study by Wang (2003) of the program's impact on production and income in Wuqi, Shaanxi, revealed that it has contributed to the improvement of pro-ductive efficiency, the increase of farmers' income, and the expansion of off-farm jobs. Dong, Zhong, & Wang (2005) found that the food security of households par-ticipating in the SLCP has been improved, compared either to the status of their own prior to implementing the program or to that of the non-participating households. Li (2004) showed that in many areas the adjustment of the rural economy, induced by the SLCP, has already benefited farmers' income growth. Given the detected posi-tive effects of labor transfer, economic adjustment, and income increase, a general implication of the above studies is that the SLCP can be sustained in the long run.

On the other hand, some researchers have questioned the effects of the SLCP on labor transfer and income increase and thus its sustainability. For instance, with household data collected in Shaanxi, Gansu, and Sichuan, Xu, Tao, & Xu (2004) found that until 2002, the SLCP had not made a significant impact on the adjust-ment of the production structure, employment in non-farming sectors, and increase in farmers' income. Using case studies in Sichuan, Guo, Gan, Li, & Luo (2005) indi-cated that because the animal husbandry was hit hard by implementing the SLCP, households participating in the program experienced a decline in their living stan-dards. Yi, Xu, & Xu (2006) also showed that while the effectiveness of the program was enhanced after 2004, its impacts on facilitating rural employment, production adjustment, and income growth remained insignificant.

Several observations can be drawn from the previous studies. First, those studies suggesting positive income and employment effects tend to focus on the direct gov-ernment subsidies that farmers have received, and the aggregate structural adjust-ment of the local economy that the program has implied. However, few have considered the induced reduction in crop and/or animal production and displacement

---

[1] This total investment is about US$32 billion, given the current exchange rate of $1 = 6.85 yuan.

of farm labor. And most of these studies lack rigor in their analyses. In contrast, those works showing insignificant or even detrimental program impacts seem to have taken a more quantitative approach as well as a more balanced and disaggregate view by incorporating the concomitant negative effects on production and employment. Moreover, they argue that without adequate government assistance and training, it is not all that easy to quickly adjust the local economy and transfer the displaced farming labor. Nonetheless, these scholars have rarely moved beyond the features of the retired lands and engaged households to account for the outcome of the program.

Also, it seems unrealistic to expect a uniform outcome of such a large program, given its broad coverage and the varying biophysical and socioeconomic conditions across the country. In addition, the location of the selected sample sites makes a difference in determining the program effects, just as the time span of an investigation does. More importantly, the effectiveness and impact of the program are predicated on the internal and external local conditions under which it is executed (Yin et al., 2008). It is thus critical to identify these conditions and incorporate them into the assessment of the SLCP impact, which is what we will do in this chapter.

Formally, the propositions we make here are that in addition to participation status and household characteristics, the impacts of the SLCP on income growth and labor transfer are determined by the local economic development, program extent, and political leadership; and the income impacts may vary from sector to sector. In other words, implementing the SLCP can result in quite different outcomes in farming, animal husbandry, and thus total income; and it is likely that the program will make a greater impact where there exist a better developed economy, a larger program extent, and a stronger political leadership.

To test these propositions, we have selected three counties – Wuqi, Dingbian, and Huachi in the Loess Plateau region covering two time periods – 1999 and 2007. While these counties are adjacent, they belong to different jurisdictions, which can better reflect the varying extents of program execution, political setting, and economic development. The time span of the study, from 1999 to 2007, represents the longest of this type of inquiry so far. Also, dividing the aggregate income into incomes from farming, animal husbandry, off-farm work, and other sources will enable us to look into the gains and losses caused to different sectors. Further, the difference in differences (DID) model that we adopt is well-suited to the task of quantifying the program's impacts on the transfer of rural surplus labor and the growth of farmers' income (Lee, 2005). So, we expect that our empirical analysis will generate a rich set of interesting results, and thus make a timely contribution to a better understanding of the program performance and a more thorough discussion of how to improve its effectiveness and impact. We also hope that our work will provide valuable information to other countries undertaking similar ecological restoration efforts.

Overall, it is found that along with other variables, participation status, local economic condition, program extent, and political leadership have indeed had significant influences on household income and off-farm employment. Moreover, the effects of participation on crop production income, animal husbandry income, and off-farm income vary substantially. These results confirm our hypotheses and have major policy implications. The chapter is organized as follows: We devote the next two sections to theory and methods, and study site and data; then, we present our empirical results in section four and our conclusions in the final section.

10.2 Theory and Methods

We hypothesize that the impacts of implementing the SLCP on income growth and labor transfer are determined by the local economic development, program extent, and political leadership, in conjunction with the participation status, and the income impacts may well vary from sector to sector. Specifically, we argue that if the program implementation involves only a small portion of the sloping farmland, its impact will be marginal; otherwise, if it covers a large proportion of the land base, then it can cause a major impact (positive or negative). Therefore, the program extent should be considered when we examine its impacts.

It is straightforward to understand the relevance of local economic condition to the program impact. In a more developed and wealthier region, not only is it unnecessary for the local cadres to profit from the program by diverting farmers' subsidies and exaggerating the set-aside targets to their own benefits, but also more local financial and personnel resources can possibly be devoted to facilitating the program implementation (Xu, Yin, Li, & Liu, 2006). Additionally, a better developed economy will provide more opportunities to absorb the displaced farm labor into off-farm and/or non-rural jobs. As a result, it is more likely for the program to take effect in increasing farmers' income and transferring farm labor (Guo & Yao, 2007). In contrast, if the local economy is such that it has little means to provide the basic administrative support, let alone supplementing the implementation and absorbing the surplus rural labor, then it will be less likely to make a difference; and it may even open up the door for the local program managers to graft part of the subsidies (Xu et al., 2004).

In addition, the program outcome is associated with the political leadership that a locale has. If the local agency is committed to its implementation, then it is more likely for the program to succeed and thus lead to a more positive impact (World Bank, 2002). Also, in a transparent political environment, it is not so hard for the farmers to track the performance of their local leaders and detect any inappropriate behavior, including corruption (World Bank, 2002). Otherwise, a non-transparent political setting makes it easy for the local agency to engage in misconduct, which can inevitably compromise the program effectiveness and constitute a disincentive to the farmers. Finally, since participating in the subsidized land conversion affects various production activities in different ways, it is expected that incomes from these activities will change dissimilarly. That is, cropland retirement can cause a yield and thus income reduction if no more improved inputs and management practices are adopted to intensify land use. In contrast, if more improved inputs and management practices are adopted, then intensified land use will not lead to a proportionate yield and thus income decrease. Also, cropland conversion and/or crop yield reduction may mean that open herding is restricted and/or feed stocks reduced, in which case income from animal husbandry can be negatively affected.

Our task in this chapter is to test the validity of the above hypotheses by fitting an adequate empirical model with a sound dataset. To that end, we have compiled repeated cross-sectional data of household production activities in three counties of the Loess Plateau region. With observations made for times both prior to and after the program initiation and for both participating and non-participating households, our DID model will allow us to detect the program impacts effectively. In particular, including variables of economic condition, program extent, and political leadership in the estimation will make it possible to explain the success or failure of the program in the proper context. To our knowledge, this is one of the first studies that have attempted to incorporate a broader set of variables, both internal and external to the program implementation, into its impact determination.

The concrete model is as follows:

$$Y_{it} = \alpha_0 + \alpha_1 T + \delta D_{it} + \beta Z_{it} + \gamma X_{it} + c_i + \mu_{it} \tag{10.1}$$

where Y is a dependent variable representing farmers' income (from different sources) or off-farm employment; i and t denote household and time, respectively; T is a time dummy, taking values of 0 for prior to the program initiation or 1 for after it; D is another dummy variable to reflect the status of program participation – taking a value of 1 if a household participates or 0 otherwise; Z_{it} represents control variables affecting farmers' income and off-farm employment, including those commonly used ones, such as family size, number of household laborers, and farmland per capita, as well as the ones that we propose to use – local program extent, economic condition, and political leadership (see discussion below); X_{it} is a group of variables that may not vary over time or may vary spontaneously, including age of the household head and a family relative serving as a village leader; c_i is a set of unobservable variables that affect family income and off-farm employment as well; and μ_{it} is the error term. Included in the parameters to be estimated are α_0, the intercept, α_1, the time effect, δ, the effect of the participation status on income growth and labor transfer, and β and γ, the effects of the control variables on the dependent variables.

Understandably, the effects of local economic condition, program extent, and political leadership on farmers' income and labor transfer are conditional on the household's engagement in the program. If so, these variables may not be directly included in Z_{it}; rather, they should enter the above equation as interactive terms with the participation dummy. We use the per capita GDP of the township to which the household belongs as a proxy for the local economic condition, the percentage of retired cropland of a household as an indication of the program extent, and another dummy variable to distinguish the political leadership of the sample counties.

After first-order differencing, the above model becomes:

$$Y_{i1} - Y_{i0} = \alpha_0 + \delta D_{it} + \beta(Z_{i1} - Z_{i0}) + (\mu_{i1} - \mu_{i0}) \tag{10.2}$$

Note that unobservable effect c_i and time invariant (or spontaneously variant) factors X_{it} have disappeared following the first-order differencing. In order to obtain consistent estimates, it is further assumed that self-choice is not a serious problem in the above model. That is, whether a household participates in the program is not an endogenous choice (Lee, 2005). Given the short time span of cropland set-aside planning and execution, this assumption seems plausible (Wooldridge, 2002). Xu et al. (2004) already demonstrated that the problem of self-choice in participation is negligible.

10.3 Site and Data

The site for this study constitutes three counties of the Loess Plateau region – Wuqi in Yan'an municipality of Shaanxi, Dingbian in Yulin municipality of Shaanxi, and Huachi in Qingyang municipality of Gansu. The rationale for this selection is the following. First, these three counties represent the typical ecological conditions found in the region, where land degradation and soil erosion were so severe that there had been a great need for farmland retirement and conversion. Second, their adjacent locations and similar landscapes as well as program implementing paths (all initiated the farmland conversion in the late 1990s and almost completely achieved the conversion by 2005) are conducive to a comparison between them. Third, their different jurisdictions make it more likely for us to capture the variations in program extent, political leadership, and economic status and thus their influences on the outcome of program implementation.

Before proceeding to presenting our data, a brief description of the basic conditions of these three counties is in order. Situated in the northeast of Yan'an, Wuqi has a total population of 127,369, of which rural residents account for 109,470. Like its neighbors, Wuqi is well known for its rich petroleum and gas reserves. But unlike its neighbors, the county has enjoyed a preferential treatment by the central government in exploiting its oil and gas reserves, which has enabled Wuqi's economy to grow rapidly in recent years. Wuqi was selected for this treatment in the mid-1980s as a result of its significance in contemporary Chinese history as the ending place of the Red Army's Long March and because of the area's extreme poverty (Wuqi SLCP Office, 2007). The county's GDP was 2.1 billion yuan in 2005, when its own revenue reached 0.7 billion yuan. Now, Wuqi has become one of the richest counties in western China (Wuqi Statistics Bureau, 2006).

Before 1998, Wuqi had a cultivated land of 123,700 ha, or 3.40 ha per household, and a large number of the rural households also raised goats, whose population peaked to 280,000. As a consequence of extensive farming and open grazing, the land and vegetation were heavily degraded, making the problems of water runoff and soil erosion extremely severe. In response, in 1998 Wuqi began retiring croplands on steep slopes and converting them to forest and grass coverage. Taking advantage of the national initiative, Wuqi's land set-aside and conversion expanded tremendously in 1999. Croplands were cut back to 10,000 ha, and open grazing was banned in favor of raising goats in pen as well as vegetation recovery (Wuqi SLCP Office,

2007). To make the ecological and economic transformation, the county government has invested heavily in such activities as improving the quality of the remaining farmland, introducing new breeds of crops and animals, and promoting best land-use practices to supplement the SLCP. Now, over 97,000 ha of converted cropland has passed the national survival, growth, and stocking inspections (Wuqi SLCP Office, 2007). Because of its decisive action and tremendous change, Wuqi has attracted broad attention. Government leaders, program managers, and journalists across the country flock there to learn its experience and lessons, and scholars from research institutions travel there to conduct field experiments and surveys.

Lying in the transitional zone between the Loess Plateau and the Erdos Desert, Dingbian is located in the west part of Yulin. Of its population of 315,851, over 87% lives in rural areas (Dingbian Statistics Bureau, 2006). Huachi is located in the eastern part of Gansu, and 86% of its 130,175 population is rural residents (Huachi Statistics Bureau, 2006). Similar to Wuqi, extensive farming and open grazing existed in these two counties. Also similar to Wuqi, these counties are endowed with rich petroleum and gas resources. However, they have not been allowed to develop these resources locally as Wuqi has. Instead, the national company, Petro China, holds the exclusive right of exploration. While figures show that the GDP of Dingbian and Huachi in 2005 was close to 3 billion yuan and 4.6 billion yuan, respectively, higher than that of Wuqi, much of that was contributed by the national oil company, which did not benefit the local treasury and employment much. So, the total budget for Dingbian and Huanchi counties was less than 60 million yuan each in 2005 (Dingbian Statistics Bureau, 2006; Huachi Statistics Bureau, 2006).

These two counties have participated in the SLCP as well. Their total amount of retired cropland is 10,966 ha for Huachi and 21,905 ha for Dingbian, suggesting a much smaller extent of program implementation given their total cropland holding of 57,265 and 83,333 ha in 1997, respectively. Also, extensive farming and open grazing in these two counties are still the norm, rather than the exception. Furthermore, their local investment in the land retirement has been negligible, and incidences of delayed subsidy delivery and even deduction of farmers' subsidies have occurred (Dingbian SLCP Office, 2007; Huachi SLCP Office, 2007). Some township officials have even attempted to use the subsidies to offset households' tax and other financial obligations.

In sum, marked differences exist between Wuqi and the other two counties. Compared to Wuqi, Dingbian and Huachi lacked the political leadership, local investment, and extensive participation. We expect that these variations will be reflected in program impacts. To capture the difference in political leadership, the dummy variable we use is 1 for Wuqi and 0 for the other two.

In August 2007, our research team conducted a survey of 200 randomly chosen households in each of the three counties, and our questionnaire included basic household characteristics, production, consumption, income, and farmland retirement and conversion. The basic characteristics of surveyed households are listed in Table 10.1. It can be seen that there is little difference in number of labor, years of average education, and age of household head between participating and nonparticipating households. Noticeable differences exist in family size, cultivated land,

Table 10.1 The Basic Features of the Surveyed Households in the Three Counties

	Non-participating Households (131)	Participating households (469)	F-test of variance	T-test family differences
Family size	4.95 (1.25)	4.63 1.51	1.46*	1.63* (0.104)
Number of laborers	2.56 (1.18)	2.45 (1.17)	1.01	0.66 (0.51)
Years of education per person	4.20 (3.67)	4.39 (4.32)	1.24	0.34 (0.73)
Age of household head	50.53 (10.73)	48.77 (10.99)	1.05	1.15 (0.25)
Years of education for household head	5.20 4.26	5.89 3.62	1.39*	1.23 (0.21)
Cultivated land	9.93 (5.29)	11.42 (7.26)	1.88**	1.66* (0.09)

Notes:

1. Of the 108 nonparticipating households, 2 in Wuqi, 62 in Dingbian, and 44 in Huachi; of the 492 participating households, 198 in Wuqi, 138 in Dingbian, and 156 in Huachi.

2. Columns 2 and 3 are the mean values for non-participating and participating households, figures in parenthesis are standard deviations; column 3 is the F test of variance uniformity of the two groups; column 4 is the t test of family characteristics.

*, and ** represent significance at the level of 10, 5, and 1%, respectively.

and years of schooling for household head, calling for their inclusion in our formal analysis.

Table 10.2 compares per capita incomes of the two household groups in Wuqi between 1999 and 2006. Except for the animal husbandry income of the participating households, all incomes increased during that period of time. The crop production income of non-participating households rose from 5,591 yuan in 1999 to 5,788 yuan in 2006, while that of participating households rose from 3,733 yuan in 1999 to 4,653 yuan in 2006. The animal husbandry income of non-participating households grew from 1,162 to 1948 yuan, but that of participating households declined from 3,575 yuan in 1999 to 1,409 yuan in 2006. The off-farm income of non-participating households rose from 2,475 to 2,917 yuan, whereas that of participating households increased from 10,404 yuan in 1999 to 13,785 yuan in 2006.

In 1999, the crop production income of non-participating households was 1,859 yuan, which was significantly higher than that of participating households. In 2006, however, this gap shrank to 1,136 yuan and became insignificant. Even though the cultivated land of participating households was greatly reduced, their improved productive efficiency could have reduced the gap of crop production income, compared to non-participating households (Chapter 13). Before the land set-aside, the two groups had significant differences in their incomes from animal husbandry, off-farm employment, and other sources as well as total income.

Table 10.2 Per Capita Income of Surveyed Households in Wuqi in 1999 and 2006

	Non-participating households		Participating households		Between group income difference	
	1999	2006	1999	2006	1999	2006
Crop production income	5,591 (7,303)	5,788 (12,417)	3,733 (3,907)	4,653 (8,860)	1,859 (2.3)*	1,136 (0.7)
Animal husbandry income	1,162 (1,734)	1,948 (3,163)	3,575 (11,951)	1,409 (1,540)	−2,413 (−2.0*)	539 (1.5)
Off-farming income	2,475 (5,711)	2,916 (7,733)	10,404 (13,867)	13,785 (24,502)	−7,930 (−5.3**)	−10,869 (−4.3**)
Other income	0 (0.0)	5,411 (3,494)	61 (603)	6,778 (8,244)	−61 (1.0)	−1,367 (−1.5)
Total income	9,228 (5,835)	16,064 (7,158)	17,773 (12,697)	26,625 (20,664)	−8,544 (−5.3**)	−10,561 (−3.4**)

Notes:

1. Crop production income is income from producing corn, potatoes, and other minor crops; animal husbandry income is income from raising livestock, predominantly goats; off-farm income is income from off-farm employment, mainly construction and service work in local towns as well as large cities; other income is income from other sources, such as family properties and government subsidies; and total income is the gross income from all sources. Note that because these statistics are rounded mean values, they may not add up to the total exactly.

2. Columns 2–5 are the mean values for the two groups, standard deviations are in parentheses; columns 6–7 are the between-group differences, the t statistic is in parentheses.

*, and ** represent significance levels of 10, 5, and 1%, respectively.

But the animal husbandry income gap narrowed and was no longer significant in 2006 due to banning open grazing, which adversely affected both groups. The difference of income from other sources between the two groups was never significant.

Table 10.3 compares incomes of the two household groups in Huachi and Dingbian between 1999 and 2006. All households witnessed an increase in their crop production income, off-farm income, income from other sources, and total income. The animal husbandry income of non-participating households dropped from 2,371 to 1,591 yuan, whereas that of participating households declined slightly. The crop production income of non-participating households increased from 2,176 yuan in 1999 to 4,511 yuan in 2006, and that of participating households also increased from 2,475 to 4,614 yuan. The off-farm income of non-participating households dropped from 6,409 to 5,568 yuan, while that of participating households rose from 6,642 yuan in 1999 to 9,912 yuan in 2006. In 1999, the crop production income of participating households was 299 yuan higher than that of non-participating households. In 2006, this gap narrowed to 104 yuan. The insignificant differences in crop production income, off-farm income, and total income between the two groups in Huachi and Dingbian indicate that their smaller share of land retirement did not make a large difference.

Table 10.3 Per Capita Income of Surveyed Households in Huachi and Dingbian in 1999 and 2006

	Non-participating households		Participating households		T-test of between-group difference	
	1999	2006	1999	2006	1999	2006
Crop production	2,176	4,511	2,475	4,615	−299	−104
income	(3,282)	(4,193)	(2,708)	(4,363)	(−0.9)	(−0.6)
Animal husbandry	2,371	1,591	1,358	1,265	1,012	326
income	(8,136)	(1,830)	(1,514)	(1,186)	(1.5)	(1.5)
Off-farm income	6,409	5,568	6,642	9,912	−234	−4,344
	(9,802)	(19,489)	(13,823)	(24,765)	(−0.1)	(−1.4)
Other income	1,459	1,708	487	535	972	1,172
	(1,355)	(5,275)	(1,020)	(1,247)	(5.8**)	(1.9*)
Total income	12,414	13,379	11,962	16,327	1,452	−2,948
	(12,661)	(1,906)	(9,703)	(12,802)	(−0.4)	(−1.9)

Notes:
1. Crop production income is income from producing corn, potatoes, and other minor crops; animal husbandry income is income from raising livestock, predominantly goats; off-farm income is income from off-farm employment, mainly construction and service work in local towns as well as large cities; other income is income from other sources, such as family properties and government subsidies; and total income is the gross income from all sources. Note that because these statistics are rounded mean values, they may not add up to the total exactly.
2. Columns 2–5 are the mean values for the two groups, standard deviations are in parentheses; columns 6–7 are the between-group differences, the t statistic is in parentheses.
∗, and ∗∗ represent significance levels of 10, 5, and 1%, respectively.

10.4 Estimated Results

Table 10.4 lists the estimated results. The goodness of fitting ranges from 0.58 to 0.25 in four of the six cases, which is encouraging for first-order differenced models. Even in the two cases (income from other sources and total income) where the R^2 is very low, it is not unusual for this type of policy, or more broadly treatment, effect assessment model (Woodridge, 2002; Lee, 2005). First, all the variables have a positive effect in the crop production income regression. Compared to that of non-participating households, the crop production income of households participating in the SLCP increases by 131.1 yuan, which is not a large figure in magnitude but significant at the 99% level. A better developed local economy, a larger program extent, and a stronger political leadership, respectively, result in an increase of the household's crop production income by 619.3, 170.2, and 251.3 yuan at the 99% significance level. Together, these add up to a sizable amount (1,240 yuan), and they have partially confirmed what we hypothesized – variations in local programmatic, economical, and political conditions all impact the crop production income. Education level of the household head also has a significant influence on crop production income, with one more year of schooling leading to an increase of 83.6 yuan. Other variables like number of household laborers, per capita cultivated area,

Table 10.4 Regression Results of Income and Off-Farm Employment, 1999–2006

	Crop production income	Animal husbandry income	Off-farm income	Other income	Off-farm employment	Total income
Status of	131.11	−2,445.52	3,170.06	382.16	0.09	5,397.04
participation	6.23	−2.67	1.54	0.14	3.05	3.87
Economic	619.27	202.64	187.94	−269.32	0.25	286.52
condition	5.90	1.04	2.63	−0.68	8.00	2.35
Program extent	170.25	73.69	62.95	−145.46	0.12	175.97
	2.57	0.63	2.63	0.05	2.15	1.97
Political	251.33	68.18	55.18	−50.79	0.07	91.63
leadership	9.08	1.14	2.16	−0.05	11.48	2.39
Education of	83.55	191.92	522.17	138.29	0.02	1,059.97
household head	67.11	1.26	1.61	1.22	1.35	2.83
Family size	8.37	507.66	191.12	1,309.85	0.14	1,867.99
	2.11	1.05	0.19	3.63	3.60	2.02
Number of	190.59	258.93	−1,792.95	−498.13	0.07	1,376.97
laborers	2.07	1.62	−1.17	−0.59	1.76	3.13
Non-agricultural	187.41	−606.91	9,191.11	126.79	–	11,046.10
employment	21.71	−1.25	5.09	0.20	–	3.44
Per capita	984.56	−159.15	−328.14	252.31	−0.02	231.62
cultivated land	2.59	−0.34	−0.33	0.69	−4.19	0.13
Intercept	−543.62	1,726.65	7,536.26	−596.58	0.49	3,052.57
	−0.18	0.99	0.94	−0.23	1.54	0.21
R^2	0.58	0.40	0.25	0.20	0.48	0.15

Note: Corresponding to each variable, the figure in first row is the estimated coefficient, and the figure in the second row is the t statistic value.

and non-agricultural employment lead to a significant increase of crop production income as well.[2]

Second, the regression of animal husbandry income reveals that participation status is negatively associated with income at the 95% significance level. The animal husbandry income of participating households is depressed by 2,445.5 yuan, in comparison to that of non-participating households. Here, program extent, economic development, and political leadership do not matter much. Variables like schooling years of household head, family size, and number of household laborers have a positive but statistically insignificant effect. Likewise, per capita cultivated area and local non-agricultural employment have a negative but statistically insignificant effect.

Third, the off-farm income is positively related to participation status and years of schooling for household head at the 90% significance level. Participation allows farmer household's off-farm income to increase by 3,170.1 yuan, and one more year of schooling for household head leads to an increase of 522.2 yuan. Local economic

---

[2] Off-farm employment includes employment in local non-agricultural activities and off-village employment as migratory workers.

development, program extent, and political leadership cause the household off-farm income to increase by, 187.9, 62.9, and 55.2 yuan, respectively. These effects are all highly significant. Additionally, non-agricultural employment has a positive effect at the 99% significance level; one more person in the non-agricultural sector results in the household's off-farm income to increase by 9,191.1 yuan. In contrast, family size, number of household laborers, and per capita cultivated area do not have strong correlations with the off-farm income. As to income from other sources, the regression has only one significant variable – family size, suggesting that the larger the family, the higher the income. All of the other variables, including the policy ones, have little effect.

Fourth, the regression of off-farm employment shows that participation has a positive effect on off-farm employment at the 95% significance level. Other things being equal, participation causes 0.09 unit of labor to shift out. Although there is a positive relation with years of schooling for household head, this relation was statistically insignificant. While family size and number of household laborers have positive effects on off-farm employment, per capita cultivated area has a negative effect on the off-farm employment. These results illustrate that: (1) the more surplus labor a family has, the more off-farm income it generates; and (2) the larger the per-person cultivated area, the less likely for the household to engage in intensive farming, making it harder to shift labor out. Local economic development has a positive relation with the off-farm employment; a coefficient of 0.25 indicates that the condition is a key factor of labor transfer. Program extent has an effect of 0.12, and political leadership has an effect of 0.07. Together, these variables cause 0.45 unit of labor to shift out of farming, which is more than four times the coefficient of participation status alone. This has further proven the hypothesis we proposed – the realized transfer of surplus farming labor depends on both the internal and external conditions, coupled with the program participation.

Fifth, the total income has a positive correlation with years of schooling for household head, family size, number of laborers, and non-agricultural employment. The contributions of these variables are 1,056 yuan from one more year of household head education, 1,870 yuan from one more person in the household, 1,377 yuan from one more family laborer, and, more substantially, 11,046 yuan from one more non-agricultural job. Participation in the land conversion program results in an increase of total income by 5,397 yuan. In addition, local economic development, program extent, and political leadership are positively correlated with the total income. Their coefficients are 287, 176, and 91.6 yuan, respectively. Again, these findings validate our basic hypothesis – the impacts of the SLCP on farmers' income are determined by local conditions, in conjunction with participation status.

10.5 Conclusions and Discussion

We set out to test the hypothesis that the impacts of implementing the SLCP are determined by the local economic conditions, program extent, and political leadership, in conjunction with participation status. We also speculated that the income effects may vary across sectors. To that end, we have estimated a difference in differences model with data collected from 600 households in three counties of the Loess Plateau region, covering both time before and after the program initiation (1999 and 2006) and both participating and non-participating categories. Our empirical results have confirmed our hypotheses nicely.

It is found that participation in the SLCP has affected incomes from different sectors in different ways. While it has a significant positive impact on crop production income, the magnitude of this effect is small. In comparison, better local economic condition, larger program extent, and stronger political leadership have much greater impacts. These results suggest that cropland retirement does not necessarily cause a reduction of cropping income if the production mode can be sufficiently transformed by adopting more improved inputs and management practices. However, participation has a substantial negative effect on income from animal husbandry, which is almost ten times the combined positive impacts of local economic condition, program extent, and political leadership. Clearly, animal husbandry was hit hard by the grazing and feeding constraints in carrying out the SLCP, even with local efforts in maintaining its vitality.

On the other hand, participation has a very large positive effect on both off-farm income and total income. In combination, these results indicate that although animal husbandry is negative affected, the program's impacts on other sectors are positive and thus more than offset the negative effects in aggregation. The results of the off-farm employment and income regressions highlight that participating in the program has accelerated the transfer of farming labor and greatly stimulated the income growth from off-farm opportunities. Moreover, these positive effects are reinforced by better economic development, larger program extent, and stronger political leadership. These findings are new to the literature, and they have provided further supportive evidence to our claim that the socioeconomic impacts of the program are indeed predicated on the local program extent and conditions, coupled with participation status. Also, they indicate that it is essential to incorporate the relevant variables into any reliable assessment of the SLCP impacts.

The government should take these elements into account in its program planning and execution. For one thing, in case it delivers great ecological benefits, the program should concentrate more on the selected sites where the local agencies are committed to an effective and transparent implementation and the local economies are conducive to intensifying cropping on reduced land, absorbing displaced surplus labor, and/or sustaining animal husbandry. But it should be made clear that the evolving local economy can alter the comparative advantages of various production and income opportunities. As such, tradeoffs between them must be weighted properly. This means that the government should identify where and by how much the production and income will contract or expand and design measures to deal with the associated winners and losers. It also implies that it may not a simple matter for the program to fulfill its dual objectives of poverty alleviation and ecological restoration.

While the findings of the negative effects of participation on animal husbandry income and the positive effect on off-farm employment and total income conform what was previously reported (Guo et al., 2005; Dong et al., 2005), the finding of a positive effect on cropping income is also new. The latter result implies that cropland reduction will not inevitably cause a crop yield and thus income decline. We conjecture that the significance of these effects has to do with the features of our sample, including selection of representative study site, coverage of a long time span, and division of total income into specific categories. It seems that in these aspects lies the distinction between our results and those of Xu et al. (2004) and Yi et al. (2006).

In addition, as an indication of family human capital accumulation, the number of schooling years of household head contributes to cropping income as well as total income. This validates the importance of education to family livelihoods (Hayami, 2003). Meanwhile, number of laborers and family size boost income from crop production, off-farm employment, and thus total income. Further, family size helps increase income from other sources, and number of laborers benefits income growth from animal husbandry. Also reasonable is the evidence that per capita cultivated land favors income from cropping and leads to less off-farm employment, which implies that while cropland retirement reduces crop production and income, it accelerates labor shift out of farming as well. Moreover, it is encouraging to observe that more favorable local conditions can work to more than offset the negative effect of land retirement on income from crop production.

Finally, it is worth noting that because the data used in this study cover only three counties in the Loess Plateau region, our findings may not apply elsewhere. To reach broader conclusions, more data should be collected from other regions. Also, follow-up analyses should be pursued to examine what will happen to the sample site of this study in the longer term.

Acknowledgments This study is sponsored by the National "11th Five-Year Plan" Science and Technology Support Project (2006 BAD03A0308), the International Research Center of Sediment, project of "Study on Wuqi County Sustainable Development" (2005-01-10) and the Key Shaanxi Provincial Forestry Special project "Ecological Forest Policy Research." The authors appreciate Professor Runsheng Yin for his careful editing.

References

Dingbian SLCP Office. (2007). *A summary report of Dingbian's implementation of the Sloping Land Conversion Program.*

Dingbian Statistics Bureau. (2006). *2005 Dingbian statistics yearbook.*

Dong, M., Zhong, B. N., & Wang, G. J. (2005). An empirical analysis of the effect of China's farmland conversion program on food security in Ningxia Hui autonomous region. *China's Population, Resources, and Environment, 1*, 104–108.

Guo, X. M., Gan, T. Y., Li, S., & Luo, H. (2005). An investigation of the Sloping Land Conversion Program implementation in Tianquan, Sichuan. *China Rural Observation, 3*, 73–79.

Guo, Z. Q., & Yao, S. B. (2007). A study of the sustainability of China's Sloping Land Conversion Program in Wuqi. *Forest Economics, 5*, 21–25.

Hayami, Y. (2003). *Development economics.* Beijing: Chinese Social Science Press.

Huachi SLCP Office. (2007). *A summary report of Huachi's implementation of the Sloping Land Conversion Program.*

Huachi Statistics Bureau. (2006). *2005 Huachi statistics yearbook.*

Lee, M. J. (2005). *Micro-econometrics for policy, program, and treatment effects.* Oxford, UK: Oxford University Press.

Li, R. N. (2004). The influence of China's land retirement efforts on rural economy in Luoyang, Henan. *Agriculture Modernization Research, 5*, 363–366.

State Forestry Administration. (2007). *2006 China forestry development report.* Beijing: China Forestry Press.

Wang, J. J. (2003). The economic and ecological foundations for converting farmland into forest and grass coverage. *Agricultural Economic Issues, 8*, 21–27.

Wooldridge, J. M. (2002). *Econometric analysis of cross section and panel data.* Cambridge, MA: The MIT Press.

World Bank. (2002). *Building institutions for markets.* Oxford: Oxford University Press.

Wuqi SLCP Office. (2007). *A summary report of Wuqi's implementation of the Sloping Land Conversion Program.*

Wuqi Statistics Bureau. (2006). *2005 Wuqi statistical yearbook.*

Xu, J. T., Tao, R., & Xu, Z. G. (2004). An empirical analysis of the cost effectiveness, structural adjustment, and long-term sustainability of the farmland conversion program in three western provinces. *Economic Research Quarterly, 4*, 139–162.

Xu, J. T., Yin, R. S., Li, Z., & Liu, C. (2006). China's ecological rehabilitation: Progress and challenges. *Ecological Economics*, 57(4), 595–607.

Yi, F. J., Xu, J. T., & Xu, Z. G. (2006). An analysis of the income and employment impacts of the farmland conversion program. *China Rural Economy*, 10, 28–36.

Yin, R. S., Yin, G. P., & Li, L. Y. (2009). Assessing China's ecological restoration programs: What's been done and what remains to be done? *Environmental Management* (submitted).

Zhi, L. (2004). Research on the multiple goal characteristics of our country's relief by grain from the backgrounds of Chinese and foreign returning farmland to forests. *Forest Economy*, 7, 29–32.

Chapter 11
An Evaluation of the Impact of the Natural Forest Protection Programme on Rural Household Livelihoods

Katrina Mullan, Andreas Kontoleon, Tim Swanson, and Shiqiu Zhang

Abstract In this chapter, we estimate the impact on local household livelihoods of the Natural Forest Protection Programme (NFPP), which is the largest logging ban programme in the world, and aims to protect watersheds and conserve natural forests. In doing so we use a series of micro-econometric techniques for policy evaluation to assess the impacts of the NFPP on two interrelated facets of household livelihoods, namely income and off farm labour supply. We find that the NFPP has had a negative impact on incomes from timber harvesting but has actually had a positive impact on total household incomes from all sources. Further, we find that off farm labour supply has increased more rapidly in NFPP areas than non-NFPP areas. This result is strongest for employment outside the village. On the basis of these results, policy implications for household livelihoods are drawn.

Keywords Natural Forest Protection Programme · Policy evaluation · Difference in differences · Propensity score matching · China · Income impacts · Off farm labour

11.1 Introduction

Assessing the impacts of ecological restoration programmes on the livelihoods of local residents is imperative for two main reasons. First, these programmes have severe equity impacts, which should be understood and mitigated when designing such policies. High biodiversity areas are often also home to poor, rural communities, while the benefits of conservation may go to wealthier residents of the same, or other, countries. In developed countries, conflicts between conservation policy and land use also occur (see, e.g., Innes, Polasky, and Tschirhart (1998) on the US Endangered Species Act), but those deriving income from the land are much more

K. Mullan (✉)
Department of Forestry and Environmental Resources, North Carolina State University, Box 8008, Raleigh NC 27697
e-mail: klmullan@ncsu.edu

R. Yin (ed.), *An Integrated Assessment of China's Ecological Restoration Programs*,
DOI 10.1007/978-90-481-2655-2_11, © Springer Science+Business Media B.V. 2009

likely to have secure property rights to that land than households in developing countries who may be affected by similar policies. The second reason for accounting for the impacts of conservation programmes on local residents is that a failure to do so may lead to conflict over the resources, reducing the potential effectiveness in environmental terms (Ferraro, 2002; Kramer, 1996).

In this chapter, we estimate the impact on local household livelihoods of the Natural Forest Protection Programme (NFPP), which is the largest logging ban programme in the world, and aims to protect watersheds and conserve natural forests. Existing studies that have attempted to evaluate the impacts of this programme have concluded that while it has been effective in reducing logging and increasing reforestation, these benefits have come at a significant cost in terms of the impacts on household livelihoods. In this chapter, we focus on evaluating the impacts of the NFPP on two important facets of the livelihoods in the affected areas, namely those of household income and employment opportunities. We argue that it is important to understand these particular impacts as the NFPP is part of a process of more general Chinese forest policy reform. This involves a shift from focusing simply on timber output, to new objectives of: (i) enhancing the role of forests in ecological rehabilitation and environmental protection; (ii) increasing timber supply through commercial investment; and (iii) promoting rural wellbeing and poverty reduction through agro-forestry (Wang et al., 2008). This means that even if the NFPP is contributing to the first objective, it cannot be considered successful if it undermines objectives (ii) and (iii) through negative impacts on household incomes and on their incentives to invest in future forest activities.

Measuring the impacts of the NFPP on household incomes and labour opportunities is not a straightforward matter, mainly because the Chinese economy has been undergoing huge changes during the period that the programme has been in place. Further, we cannot observe household income and labour decisions both in the presence and the absence of the programme and therefore face an identification problem. To address this, we treat the NFPP as a natural experiment, using panel survey data to compare the changes in income and labour opportunities over time in the areas where the programme was in place with changes in the areas where it was not introduced. In order to ensure the robustness of our results, we estimate the changes in income resulting from the programme using various parametric and semi-parametric policy evaluation techniques and we find that the NFPP has had a negative impact on incomes from timber harvesting. However, it has actually had a positive impact on total household incomes from all sources. Because the increase in total income appears to be driven by an increase in income from off-farm employment, we also estimate the impact of the NFPP on household participation in off-farm labour markets. We predict that the reduction in the marginal return to labour in forest activities resulting from the ban on logging would lead to a reallocation of labour from forestry to off-farm occupations. Using similar evaluation methods as for the income impacts, we estimate the effect on off-farm employment and find that it has increased more rapidly in NFPP areas than non-NFPP areas. This result is strongest for employment outside the village. We note that the increase in total household income may not correspond to an increase in household welfare because if that were the case, non-NFPP households would be expected to participate in the off-farm labour market to a greater extent as well.

11.2 Background to the NFPP

11.2.1 Description of Programme

The Natural Forest Protection Programme (NFPP) was introduced by the Chinese government in response to the serious drought in the Yellow River area in 1997 and the flooding of the Yangtze River in 1998. Both of these events were understood to have been brought about by poor agricultural practices and the removal of natural forests. The Sloping Land Conversion Programme (SLCP) was intended to address the former problem, while the NFPP addressed the latter. The NFPP applies to 17 provinces and autonomous regions, primarily in the upper reaches of the Yangtze River, the upper and middle reaches of the Yellow River. These provinces contain 73 million ha of natural forests, which amount to 69% of the total natural forest area in China (Swanson et al., 2006).

The objectives of the programme are: to restore natural forests in ecologically sensitive areas, protecting and enhancing biodiversity; to plant forests for soil and water protection; to increase timber production in forest plantations to meet national demands for timber and contribute to economic development of rural areas; and to protect existing natural forests from excessive cutting (Zhang et al., 2000). These objectives are being pursued with the use of, firstly, a ban on logging in natural forests across many parts of the country; and secondly, measures to encourage the development of new plantation forests which should increase timber supplies while reducing pressure on natural forests. These activities have been supported with funds from the Chinese government, directed specifically at: (i) afforestation and forest protection, including mountain closure, tree planting, construction of sapling bases, payments for forest tending and forest fire prevention; (ii) compensation for unemployed state forest workers, retirement pensions for state forest staff and some compensation to local governments for losses of forest taxation revenues (Xu, Katsigris, & White, 2002).

11.2.2 Existing Studies

The NFPP is the largest programme of its kind in the world, and therefore there is considerable interest in its impacts. The most significant study so far has been carried out by Xu et al. (2002), who found that while there have been positive impacts on forest cover, there have been certain negative effects such as losses in employment in the state forest sector; reductions in local government finances; and losses of employment and income for households dependent on the state and collective forest sectors. Other studies have largely been case studies of forest management by households within a small number of villages or townships, for example Shen (2001), Weyerhaeuser et al. (2006), and Demurger and Fournier (2003).

11.2.2.1 General Findings

The NFPP is largely accepted to have reduced timber harvesting in natural forest areas (Xu et al., 2002; Demurger & Fournier, 2003). As it has been implemented in important watersheds of the major rivers of China, and in areas of high biodiversity, this is expected to have positive environmental impacts, in particular on soil erosion and water conservation (Yang, 2001). However, two key environmental issues have been identified: the first is that there are suggestions from case studies that the programme has reduced the incentives to actively manage forest land by monitoring against illegal logging and forest fires and by thinning and maintaining the timber stands. This is particularly likely where the ban has been applied to village plantations (Miao & West, 2004; Weyerhaeuser et al., 2006). The second issue is that the reduction in timber supply within China is expected to have negative effects on neighbouring countries (Weyerhaeuser et al., 2006). This raises concerns because much of the imported timber originates in Asia-Pacific countries where unsustainable harvest practices and illegal logging are common (Katsigris et al., 2004).

Much of the research into the impacts of the NFPP has focussed on the impact on the state forest sector as this was initially targeted in the programme (Zuo, 2002). During the early stages, it was estimated that around 1 million employees would have to be laid off (Yang, 2001). Many of these have been re-employed in forest stewardship, while those who were not re-employed were offered compensation of three times their average salary for the previous year (Yang, 2001). While there are some cases in which compensation has not been paid in full, and where state enterprises are unable to pay full pensions for retired workers, greater losses in state forest areas have been experienced by those who were not formally employed by the state sector (Katsigris, 2002). There have also been impacts on the provision of public services by local governments as their revenues have dropped significantly in areas where forestry played an important role in the local economy (Katsigris, 2002). For example, in some counties in Sichuan Province, timber revenues represented more than 80% of total income from taxation; as timber production falls, these revenues will fall accordingly (Yang, 2001).

Yang (2001) also highlights economic benefits from the NFPP. He argues that the 1998 floods affected hundreds of millions of people and caused extensive damage, and that the result of this was major expenditures on flood defences against future disasters, including ¥ 7.8 billion on embankments along the Yangtze. The financial and social costs of both the floods themselves and of protection against future events are real costs that should be included in an evaluation of the programme. A further benefit he suggests is diversification of economic activities, giving the example of Aba Prefecture which has promoted tourism as an alternative to logging. The prefecture experienced increases of around 130% in both numbers of tourists and tourist revenues between 1997 and 1998 and now receives 30% of GDP from tourism.

11.2.2.2 Findings Relating to Collective Forests

The NFPP was primarily targeted at state forest areas, but in many places it has been extended to cover collective forests as well (Zuo, 2002). A number of studies have argued that where households and communities are prevented from accessing collective forest land, this infringes their existing rights to use that land, and their ownership of the trees on the land (e.g., Shen, 2001; Katsigris, 2002). There is little in the way of quantified economic impacts on collective forest communities. Katsigris (2002) looks just at Sichuan Province, and finds that timber output from collective forests had fallen to 6% of previous levels by 2000, resulting in revenue losses of approximately ¥ 1 billion for the whole sector. She also finds a 65% reduction in the number of forestry related township enterprises, with corresponding reductions in employment and output of 53% and 30%, respectively. The same study finds losses in tax revenues, decreases in employment, and reductions in overall economic growth rates in some counties with collective forest areas, although those with alternative development opportunities were observed to recover quickly from 2000 onwards.

Chen et al. (2001) measure income changes for three townships in Li County, Sichuan Province. In one of the townships, they observe a 17.6% reduction in per capita income between 1998 and 2000. In the other two townships, per capita incomes fall by around 1/3. These reductions are explained by losses of employment in the timber harvesting or processing sectors, or loss of revenues from provision of services to temporary migrants who were previously working in the forest sector. Shen (2001) studies two villages in Aba Prefecture, Sichuan Province and finds that households have not only lost income from the timber sector, but also through loss of access to non-timber forest products, including fuelwood, and access to grazing land for livestock. In general, the NFPP does not prohibit the collection of non-timber products, but in some cases it has either been disallowed by the local government or access has been prevented by the closure of transport routes to forest areas (Zuo, 2002).

11.2.3 Contribution of this Study

Despite the number of studies that have been carried out into the impacts of the NFPP, there has been an absence of analysis using quantitative information to assess these impacts, as those cited have relied on qualitative case study information. Case studies, such as those carried out by Xu et al. (2002), Shen (2001), Weyerhaeuser et al. (2006), and Demurger and Fournier (2003), were each focused on a small number of villages, and based on interviews with key members of state forest enterprises and forest communities. Their conclusions have raised considerable concerns about the NFPP but have as yet not been investigated more widely. Therefore, this chapter attempts to evaluate the impacts of the programme and revisit some of the conclusions of the existing studies, using a survey of 285 households in 40 villages in the south of Guizhou Province. The existing case studies are largely based in the south-western region of Sichuan and Yunnan Provinces. So far there has been no research into the impacts in the southern collective forest areas. As the forest sector in the southwest is rather different in character to the forest areas in the south of China (Rozelle et al., 2000), data from Guizhou will be the only assessment of the impacts of the ban in the southern provinces. Finally, the existing studies of the NFPP are based on information that was collected shortly after the logging ban was introduced, so this chapter supplements those findings by using data from 2004, when some adjustment to the ban has taken place and the medium term impacts are more apparent.

The findings of the existing studies, in particular that the NFPP has resulted in losses of employment and income for rural households, are concerning for a number of reasons. The first is that as discussed above, current forest policy in China is aimed in part at improving standards of living in forest areas so if the NFPP is having the opposite effect, the overall objectives may not be met. In addition, the southern, south-western and north-eastern forest areas of China, where the NFPP has been applied, are remote from the major centres of economic activity and growth and contain the poorest provinces of China. Even within these provinces, the forest locations are largely in inaccessible mountain areas, where alternative income generating options are limited. Therefore, if the NFPP reduces incomes in these areas, it will contribute to increasing poverty in already poor regions and will exacerbate the growing inequality between these regions and the rapidly growing urban centres in other parts of China.

This chapter focuses specifically on the impacts of the NFPP in collective forest areas. There are two reasons why these areas are of particular interest and concern. The first is that, unlike in the state forest areas, where the employees of the forest sector have received compensation for lost income, there has been no compensation for households in collective forest areas. The second reason is that, as argued by Shen (2001), the NFPP has infringed on the rights to forest land that were allocated to households in the early 1980s, and this has occurred without any compensation for those households. We use data from collective forest areas to examine the impacts of the programme and whether compensation should have been paid for infringements on property rights.

11.3 The Policy Evaluation Problem

The objective of this chapter is to find out the impact of the NFPP on household incomes and off farm labour decisions in affected areas. In order to do so, we take advantage of the growing literature on econometric methods for policy evaluation, for example, Heckman and Robb (1984), Ashenfelter and Card (1985), Heckman, Ichimura and Todd (1997) etc. All of these methods deal with the identification problem faced when attempting to determine the impacts of a policy intervention: if there are two potential outcomes, Y_1, the outcome when the individual participates in a programme; and Y_0, the outcome when the individual does not participate, then the impact of participating in the programme is simply given by:

$$\Delta = Y_1 - Y_0 \tag{11.1}$$

However, estimating this requires information on both Y_1 and Y_0 for each individual, which is not obtainable because we cannot observe the outcome of participation for non-participants or the outcome of non-participation for participants. Therefore, estimation of the causal effect of the programme is equivalent to solving a missing data problem (Heckman et al., 1997), and requires the use of techniques that allow the identification of the relevant impacts in the absence of the data. In the case of the NFPP, we want to know the difference between the level of household income and labour supply when a household is in the programme and when it is not, but we only have data on one situation or the other.

For any individual, the observed outcome (Y), following Roy (1951), Quandt (1972) and Rubin (1978), is defined as:

$$Y = DY_1 + (1 - D)Y_0 \qquad (11.2)$$

where D denotes participation in the programme, and takes the values 1 (if the individual participates/is treated) or 0 (if the individual does not participate/is not treated). This is commonly written as a function of observables (X) and unobservables (U_1, U_0):

$$Y_1 = g_1(X) + U_1 \qquad (11.3)$$

$$Y_0 = g_0(X) + U_0$$

The literature mentioned above uses various methods to estimate a counterfactual outcome against which the outcome for treated individuals can be compared. In the social sciences where truly randomised policy experimental data is hard to find, alternative methods have been developed that account for the fact that those participating in a programme will have different expected outcomes from those not participating. This is either because they choose to participate in the programme on the basis of their expected returns from it; or as in the case of the NFPP, because the programme affects all individuals in a particular area, or of a particular segment of the population, so the control group must necessarily have different observable or unobservable characteristics and therefore different expected outcomes.

For the purposes of this chapter, in order to determine the impact of the NFPP on household incomes and labour decisions, we will use three identification methods: difference-in-differences (DID) with covariates; Heckman et al. (1997) propensity-score matched difference-in-differences (PSM); and Abadie's (2005) propensity score weighted difference-in-differences (PSW). Those living in the counties where the programme was implemented are considered to be 'treated', while those living in the areas where it was not implemented are 'untreated'. All of these approaches allow us to account for both observable and unobservable variation between households in the NFPP and non-NFPP areas, but comparison of the results from the three methods will also indicate how sensitive those results are to the specification of the functional form of the econometric model. We first provide a brief overview of these methods before we turn to the discussing the data used and the econometric results.

11.3.1 Difference in Differences

The NFPP can be viewed as a natural experiment, with data available for a period before the policy reform and a period after the reform, and also data for two counties where it was implemented and one similar county where it was not implemented. Therefore, a relevant method of evaluation would appear to be the (DID) estimator used by Ashenfelter and Card (1985). They use a components of variance framework in which the unobservable variation is decomposed as follows:

$$U_{it} = \phi_i + \theta_t + \mu_{it} \qquad (11.4)$$

where ϕ_I is an individual effect that remains constant over time; θ_t is a time specific effect that varies over time but is the same for all individuals (e.g., common macroeconomic impact); and μ_{it} is a temporary individual specific effect that varies across time and across individuals. The relationship between treatment and outcomes is therefore given by:

$$Y_{i,t} = \alpha D_{i,t} + \phi_i + \theta_t + \mu_{it} \qquad (11.5)$$

If D is independent of μ_{it} then comparing pre- and post-treatment outcomes for participants and controls allows the identification of α, the treatment effect.

The DID estimator compares the changes in outcomes for the group of treated individuals with changes in outcomes for a control group. The basis for the estimator is that trends in the outcome variable over time, and time-invariant individual specific variation, are cancelled out. The former is based on the assumption of common trends, in other words, that the group participating in the programme would have experienced the same change in the outcome variable between time period t_0 (before the programme) and t_1 (after the introduction of the programme) as those not participating. The validity of this assumption will be discussed further below, but if it is accepted, then the difference between the change in the outcome variable for the participating group and the change in the outcome variable for the control group will give an estimate of the impact of the programme:

$$\alpha_{DID} = E(Y_{i1} - Y_{i0}|D{=}1) - E(Y_{i1} - Y_{i0}|D{=}0) \qquad (11.6)$$

An advantage of using DID is that it removes a significant part of unobservable variation that may affect outcomes regardless of the effect of the programme or policy. However, if participation is related to μ_{it}, then it will not be possible to estimate the treatment effects separately from the temporary individual specific effects. A commonly quoted example of this is the 'Ashenfelter dip' (Ashenfelter, 1978) whereby an individual is more likely to enter a training programme if they experience a temporary dip in earnings just before the training programme is introduced, indicating a relationship between individual transitory shocks that affect pre-treatment earnings and programme participation. A more serious problem in the context of the NFPP is that the results will be biased if common macroeconomic trends have different impacts on the treatment and control groups due to their observable or unobservable characteristics (Blundell & Costa Dias, 2000).

In order to deal with this particular problem, Abadie (2005) suggests that one way to control for observed characteristics that could affect the dynamics of the outcome variable is to include a vector of these characteristics in the difference-in-difference model. They may be introduced in a linear fashion so that the model becomes:

$$Y_{it} = \alpha D_{i,t} + X\beta + \phi_i + \theta_t + \mu_{it} \tag{11.7}$$

and the treatment effect becomes:

$$\alpha_{DID} = E(Y_{i1} - Y_{i0}|X, D=1) - E(Y_{i1} - Y_{i0}|D=0) \tag{11.8}$$

Alternatively, the variables may be included using interaction terms, which allows for heterogeneous treatment effects across individuals.

11.3.2 Propensity Score Matching

A drawback of the DID model with covariates to control for observable characteristics is that a particular functional form must be imposed on the model in order to estimate the parameters. In order to overcome this, two methods of semi-parametric DID estimation have been suggested. The first method was developed by Heckman et al. (1997). They use PSM to create a control sample with the same observable characteristics as the treatment sample. If the dynamics of the outcome variable are based on these observable characteristics, then, as with the use of covariates within the OLS framework, this would mean that the common trends assumption of DID could be accepted.

The basis of the Heckman et al. (1997) method is that, conditional on X, (Y_1, Y_0) and D are independent.

$$(Y_1, Y_0) \perp D|X^1 \tag{11.9}$$

This means that if we condition on observable characteristics, non-participant outcomes have the same distribution that the participants would have experienced if they had not participated in the programme:

$$F_{(y0} | X, D=1) = F_{(y0} | X, D=0) \tag{11.10}$$

A further requirement is that $0 < Pr\ (D=1|X) < 1$. If these assumptions hold, then matching can be used to construct a control sample from the outcomes of

---

[1] In fact, only the assumption that $Y_0 \perp D|X$ is necessary to estimate the average treatment effect on the treated using matching techniques.

non-participants that would be equivalent to the control group in a random experiment. The conditional independence assumption underlying the matching method is a strong one as it assumes that all selection into the programme is on observable characteristics. In reality, it may be that a set of observable variables for which the condition holds do not exist. In other words, selection may occur on unobservables. The second assumption ensures that a common support region exists. If there is a combination of X variables for which $Pr(D=1)=1$, then it is not possible to construct a control sample with the same characteristics from the non-participant observations.

The PSM (Q) method generates a control sample by using the outcome for a single untreated individual (j), or weighted group of untreated individuals, to represent the outcome that would have been attained by a treated individual (i) if they had not been treated. The comparison individual, or group of individuals, is selected on the basis of the similarity of their observable characteristics to each treated individual (i).

$$\text{ATT} = \sum_{i \in \{D=1\}} W_{N0, N1}(i) \left[Q_{1i} - \sum_{j \in \{D=0\}} W_{N0, N1(i, j)} Q_{0j} \right] \qquad (11.11)$$

Each participant may be matched with a single non-participant, as in the case of nearest-neighbour matching, or they may be matched with a weighted average of some or all of the non-participant observations, using, e.g., kernel or local linear regression.

Rosenbaum and Rubin (1983) have shown that if it is possible to match on values of X, it is also possible to match on a function of X. They use the propensity score, which is an estimate of the probability that an individual is in the treated group, given their values of X.

$$P(X) = Pr(D=1 \mid X) \qquad (11.12)$$

This can be estimated using a probit or logit model. Matching is then carried out on the individuals with the most similar propensity scores.

11.3.3 Extensions to Matching and DID

PSM has been widely used as a way to estimate counterfactual outcomes for participants in various policies and programmes. However, Heckman et al. (1997) extend the original framework of Rosenbaum and Rubin (1983) in two ways. The first is to match using panel data, generating a semi-parametric conditional DID estimator. If the following assumption holds:

$$E(Y_1^0 - Y_0^0 | X, D=1) = E(Y_1^0 - Y_0^0 | X, D=0) \qquad (11.13)$$

where Y_1^0 is the outcome in the post-treatment periods in the absence of treatment and Y_0^0 is the outcome in the pre-treatment period in the absence of treatment, then the average treatment effect on the treated is:

$$ATT = E(Y_1^1 - Y_0^0|X, D=1) - E(Y_1^1 - Y_0^0|X, D=0) \qquad (11.14)$$

This has the advantage that we can control for both observable differences between the two groups that affect the outcome variable, and any unobservable differences that remain constant over time or are related to the observable characteristics of the individual (Abadie, 2005). For this method, PSM is carried out as described above, but instead of using $Q_{1i} = Y_{1i}$ and $Q_{0j} = Y_{0j}$ to estimate the matched treatment effect, we set $Q_{1i} = (Y_{1it} - Y_{0it})$ and $Q_{0j} = (Y_{0jt} - Y_{0jt})$.

Abadie (2005), following Hirano, Imbens and Ridder (2001), proposes PSW as an alternative method to balance the treatment and control samples with respect to observable characteristics before carrying out DID estimation of treatment effects. This is done by weighting the control observations on the basis of the similarity of their propensity score to the propensity scores of the treatment group so that the untreated observations that have similar observable characteristics are weighted more than those that are very different. The result is that the same distribution of the observable characteristics is imposed on the treated and control samples. As with other DID models, this version assumes that:

$$E(Y_1^0 - Y_0^0|X, D=1) = E(Y_1^0 - Y_0^0|X, D=0) \qquad (11.15)$$

i.e. that conditional on observables, the treated and control samples would have experienced the same outcome dynamics in the absence of treatment. Also in line with the other models, we assume that the support of the propensity score for the treated sample is a subset of the support of the propensity score for the control sample. The basis of this model is:

$$E[Y_1^1 - Y_1^0|X, D=1] = E[\rho_0(Y_1 - Y_0|X)] \quad \text{where} \qquad (11.16)$$

$$\rho_0 = \frac{D - P(D = 1|X)}{P(D = 1|X)(1 - P(D = 1|X)} \qquad (11.17)$$

The average treatment effect on the treated is then given by:

$$ATT = E\left[\frac{Y_1 - Y_0}{P(D = 1)} \cdot \frac{D - P(D = 1|X)}{1 - P(D = 1)|X)}\right] \qquad (11.18)$$

Abadie (2005) describes how this imposes the same distribution of covariates on the treated and untreated samples by weighting down the distribution of $Y_1 - Y_0$ for the untreated whose values of the covariates are over-represented, and weighting up $Y_1 - Y_0$ for the observations whose values of the covariates are under-represented.

11.4 Impact of NFPP on Household Incomes

11.4.1 Econometric Methods

The impact of the NFPP on incomes will be estimated using three different methods. The methods all take advantage of the panel data available by differencing out the time-invariant unobservable heterogeneity between the participating and non-participating households. However, they use different approaches to controlling for observed heterogeneity.

11.4.2 Descriptive Statistics

Using the methods described, we estimate the impact of the NFPP on net income per household; net income per head; and income per household from timber, forest products, and employment. Table 11.1 gives the values of these variables in 1997 and 2004, for all households and separately for NFPP and non-NFPP households. All the income values include income in cash and in kind, i.e., products harvested by the household for their own use. Where crops, livestock or forest products are consumed by the household, we have used imputed prices based on the average received by those households that have sold the product.

From the descriptive statistics, we can see that total household incomes have risen across both the treated and control groups between 1997 and 2004, as have incomes from employment. Over the same period, incomes from forest products have remained approximately constant and incomes from timber have fallen for NFPP households and risen slightly for non-NFPP households. Clearly, changes in

Table 11.1 Descriptive Statistics for Dependent Variables

Income measure		1997		2004	
		Mean	S.D.	Mean	S.D.
Total income per	Total	6,933	8,846	13,258	12,697
houshold	NFPP	6,960	6,598	14,109	13,863
	non-NFPP	6,877	12,046	11,473	12,343
Total income per	Total	1,390	1,504	2,593	2,322
head	NFPP	1,458	1,507	2,822	2,387
	non-NFPP	1,246	1,494	2,113	2,114
Income from timber	Total	225	1,158	85	331
	NFPP	275	1,256	64	263
	non-NFPP	118	919	129	441
Income from	Total	713	1,314	732	1,346
non-timber forest	NFPP	586	876	533	874
products	non-NFPP	981	1,913	1,151	1,945
Income from	Total	3,040	7,120	8,130	11,736
employment	NFPP	3,196	5,776	9,089	11,394
	non-NFPP	2,715	9,364	6,117	12,244

All values in ¥ per year

the magnitude and structures of incomes have been occurring, some of which will be due to the NFPP and some of which will be due to wider changes across the rural economy. Because of this, it is necessary to isolate the impacts of the NFPP from other types of variation. As discussed above, we do this by controlling for both observable and unobservable factors that are expected to contribute to changes in income from different sources, and that may vary between the treated and control samples. This is done either through the use of covariates in the DID models, or by estimating a propensity score for participation in the NFPP which is then used to weight the observations for estimating the impacts of the NFPP on income.

The same observable variables are used for the estimation of the propensity score and as covariates in the parametric DID model. These are shown in Table 11.2. Following the recommendations of Caliendo and Kopeining (2006), only variables that may affect both participation and outcomes are included, and variables that are potentially affected by participation or the anticipation of participation are excluded. Heckman et al. (1997) also stress that the data for the treated and control groups should come from the same survey, which holds for the dataset used in this study.

11.4.3 Propensity Score

For the matching and weighting methods, the propensity score was estimated with a probit model, using the same variables as the controls in the parametric estimation. The results of the probit estimation are shown in Table 11.3, and the distribution of the scores for NFPP participants and non-participants is shown in Figs. 11.1 and 11.2.

These results show that at a 5% level of significance, households in NFPP areas are more likely to be from the relatively prominent Dong or Han ethnic groups than smaller minority groups, and they have higher levels of education; NFPP households tend to have fewer adults, but more forest land, and are relatively remote from their county town.

These densities show that the households in the NFPP areas have higher propensity scores than those outside the NFPP areas, which is as we would expect. An impact of this is that in the matching process, the NFPP households with higher propensity scores will be compared with a relatively small number of control observations. However, there are at least some control observations with high propensity scores, and related to this, there is a large area of common support on which to estimate the impacts of the policy.

The balancing properties of the propensity score were estimated using the 'blocking' method of Dehejia and Wahba (1999). The purpose of this is to ensure that the propensity score has effectively balanced the distribution of the covariates in the treated and control groups. More specifically, we test whether, after conditioning on the propensity score, treatment is independent of the observable covariates, X.

$$D \perp X \mid p(X) \qquad (11.19)$$

The method of Dehejia and Wahba (1999) investigates an approximation of whether $f(X|D = 1) = f(X|D = 0)$ by dividing the sample into blocks based on

Table 11.2 Descriptive Statistics for Independent Variables

Variable		Mean	Median
Ethnic group dummy – Miao	Total	0.29	0.45
	NFPP	0.23	0.42
	non-NFPP	0.40	0.49
Ethnic group dummy – Han	Total	0.15	0.36
	NFPP	0.18	0.39
	non-NFPP	0.09	0.28
Ethnic group dummy – Dong	Total	0.51	0.50
	NFPP	0.57	0.49
	non-NFPP	0.38	0.50
Whether HH head has more than	Total	0.42	0.49
primary education (Dummy var)	NFPP	0.45	0.50
	non-NFPP	0.35	0.48
Age of HH head (years)	Total	47.66	11.31
	NFPP	48.77	10.80
	non-NFPP	45.32	12.02
Number of adults in the household	Total	2.95	1.17
(number)	NFPP	2.88	1.12
	non-NFPP	3.10	1.27
Share of income from forests (%)	Total	0.14	0.14
	NFPP	0.12	0.14
	non-NFPP	0.16	0.16
Area of household forest land (mu)	Total	37.89	105.82
	NFPP	48.26	123.39
	non-NFPP	16.13	6.03
Distance from village to county town	Total	29.67	14.06
(in km, as crow flies)	NFPP	30.67	15.09
	non-NFPP	27.64	11.43

Table 11.3 Estimation of Propensity Score

Variable	Coefficient	Standard errors	P value
Dong	1.450	0.386	0.000
Miao	0.686	0.398	0.085
Han	1.654	0.449	0.000
Age	0.015	0.008	0.063
Education	0.490	0.179	0.006
No. of adults	−0.142	0.074	0.054
Share of income from forests	−0.705	0.579	0.223
Area of forest land	0.006	0.002	0.001
Distance from village to county town	0.018	0.006	0.006
Intercept	−1.738	0.591	0.003
Log likelihood	−147.95		
Pseudo R^2	0.175		
Number of observations	285		

Fig. 11.1 Propensity Scores
for Non-NFPP Households

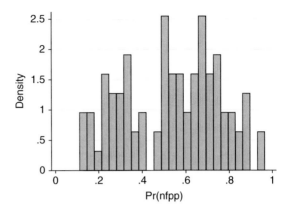

Fig. 11.2 Propensity Scores
for NFPP Households

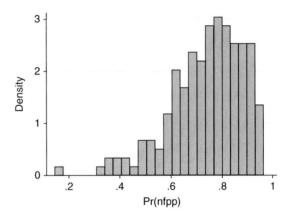

the estimated propensity score, and testing whether the means and standard deviations of each covariate in each block are equal. Using the STATA programme *pscore* (Becker & Ichino, 2002), we find that at a 1% level of significance, the treated and control groups are balanced for all covariates, in all blocks.

11.5 Results

In order to test the impact of the NFPP on household incomes from different sources, we estimate DID models, with and without covariates, a PSW model, and carry out PSM, using kernel and local linear regression. All the models are estimated on changes in income between 1997 and 2004, with the covariates described in the previous section.[2] The results are shown in Table 11.4.

---

[2] All estimation is carried out in STATA 9. The matching estimates use an Epanechnikov kernel for the kernel matching process, and a bandwidth of 0.1 for both the kernel and local linear matching. Alternative matching specifications were also estimated to assess robustness, and these did not

Table 11.4 Impacts of NFPP on Income from Different Sources

	DID no covariates	DID with covariates	PS matching, kernel regression	PS matching, local linear regression	PS weighting
Total income per HH	2,694	3,043	2,871	2,794	1,141
	(1,337)**	(1,502)**	(1,590)*	(1,529)*	(2,917)
Total income per capita	510.8	466.4	468.9	442.3	32.04
	(269.6)*	(308.4)	(326.3)	(336.5)	(702.6)
Income from timber per HH	−224.5	−270.1	−266.2	−313.0	−498.9
	(157.7)	(162.3)*	(109.9)**	(123.2)**	(275.2)*
Income from NTFPs per HH	−227.7	−37.65	−31.01	−27.38	−104.3
	(98.9)**	(106.44)	(183.2)	(180.5)	(155.9)
Income from employment per HH	2,548	2,399	1,965	2,019	1,031.8
	(1,293)*	(1,443)*	(1,423)	(1,351)	(2,140)

* significant at 10%
** significant at 5%

Looking first at the results of the DID model without any covariates, we find that the NFPP had no significant impact on income from timber. However, this specification suggests significant negative effects on income from non-timber forest products, and significant positive effects on income from employment and total income. This model controls for unobservable variation that has the same impact on both treated and control groups over time. However, it does not control for observable variation between the two samples that may affect the impact that general trends have on the two groups. For example, general trends that were occurring between 1997 and 2004, such as increasing off-farm employment, rapid income growth, and a switch away from fuelwood (a key non-timber product) towards other fuels, may have affected different households in different ways. If these are related to observable characteristics of the household, the inclusion of covariates or use of the propensity score should control for them, and therefore better isolate the impacts of the NFPP from other macroeconomic changes.

This is supported by examination of the results of the remaining specifications, all of which incorporate variation in observable characteristics. The results obtained using DID with covariates, PSM, and PSW give broadly similar conclusions in terms of the direction and significance of the impacts of the NFPP. In all of the conditional specifications, participation in the NFPP is found to have a significant negative impact on household income from timber. Based on the estimates of the DID model with covariates and the PSM models, households in the NFPP experienced a reduction in timber income of around ¥ 300 between 1997 and 2004, relative to

significantly affect the results. The PSW procedure was carried out by weighting each observation by the propensity score, and then calculating the average of these weighted estimates. The standard errors for the PSM and PSW methods are estimated using bootstrapping with 500 repetitions. Lastly, all the models were estimated only within the region of common support.

what they would have experienced if they had not participated in the programme. The PSW model suggests a greater impact on timber incomes, of around ¥ 500.

The impact of the NFPP on non-timber forest products is negative in all of the models, but small and not significantly different from zero. We would not expect a significant negative effect on non-timber products as collection is still permitted in our survey areas. Regarding fuelwood, although its collection is still allowed under the NFPP, there has been some reduction in quantities collected in both NFPP area and non-NFPP areas which is due to increasing use of alternative forms of energy such as biogas and electricity. It is possible that the NFPP could increase the availability of certain non-timber products in the longer term, but we would not expect to observe this type of impact at this stage.

All of the specifications used show a positive and significant impact of the NFPP on total household incomes. The increase in total household incomes amounts to around 3,000 ¥ on average in the DID and matching models, while the PSW again gives somewhat different results, with a lower impact on total incomes, which is not significantly different from zero. The results suggest that per capita incomes have increased by approximately ¥ 450 between 1997 and 2004, although this effect is not significant at 10%. Possible explanations for the finding that incomes have increased as a result of the NFPP will be discussed in the following section.

Finally, income from employment has increased by around ¥ 1,000 more in NFPP areas than non-NFPP areas according to the PSW model, while the other models suggest a difference in the change in income from employment of ¥ 2,000 between the treated and control groups. However, this effect is not significantly different from zero, apart from in the DID model where the significance is border-line at 10%.

11.5.1 Impacts of NFPP on Off-Farm Employment

11.5.1.1 Theoretical Impacts

The results of the previous section indicated that the NFPP has resulted in a drop in income from timber, but an average increase in total income. The first of these effects is in line with a logging ban policy. However, the positive effect on total incomes appears to be somewhat counterintuitive because we might expect households affected by the NFPP to be worse off overall. In fact, previous studies on the impacts of the NFPP have suggested that this is the case, although largely not on the basis of quantitative information. As income from off-farm employment has increased faster in NFPP areas than non-NFPP areas, while income from other sources has declined or remained unchanged, it appears that the change in total income is primarily driven by changes in income from employment. We therefore examine the effect that the NFPP has had on household participation in off-farm employment. In particular, we consider how falling incomes from timber would affect the labour allocations decisions of households who do potentially have access to off-farm labour markets, but were not previously choosing to participate in them.

Following the standard model of farm household decision making (e.g. Singh, Squire, & Strauss, 1986) households affected by the NFPP will allocate labour to forest activities, agricultural production and off-farm labour until the marginal products of labour in forest activity and agriculture are equal to the wage rate in off-farm employment. If the return to labour in forest production falls as a result of the ban on logging, the total household labour allocation to forest activities will be reduced to the point where the marginal product of labour again equals the off-farm wage rate. Under standard assumptions that the agricultural production function exhibits diminishing returns to land and labour and the wage rate in the labour market is constant, we would expect that labour would move from forest production into off-farm employment rather than into agriculture. In this case we would expect the NFPP to increase participation in the off-farm labour market.

11.5.1.2 Empirical Results

We now empirically evaluate the impact of the NFPP on off-farm employment. We estimate the impact on total off-farm employment, and also look separately at work that requires household members to migrate away from the village on a temporary or permanent basis. The amount of employment is measured as the number of household members participating, and as the total number of days spent in a particular activity. The same estimation methods are used as for the previous income impacts evaluation, and the variables used to estimate the propensity score are also the same.

Looking at the results in Table 11.5, the impact of the NFPP on the numbers of household members working off-farm is positive but not significantly different from zero. The impact on the total number of days worked off-farm by all household members is also positive. This impact is significant in the DID models with and without covariates, but not in the PSM or PSW models. Across all the models, the average impact of the NFPP on total days worked off-farm ranges between 50 and 100 days per household.

The impact on the number of household members leaving the village for employment purposes is again positive, with mixed evidence on the significance of the effect, although it is borderline at a 10% significance level in the PS-based models. Finally, the results indicate that households in NFPP areas increased the number of days worked outside the village by an extra 80–130 days, compared with households in non-NFPP areas. This finding is significant in all except the PSW model, which is more conservative across all the specifications.

Overall, the results suggest that, while there may be some positive impact of the NFPP on overall employment, the evidence for it is not strong. In contrast, there is a clearer positive impact on the number of days that household members spend as migrant workers. Based on the discussion in the previous section, these findings indicate that the extent to which households can increase their participation in local labour markets may be more limited than the extent to which they can increase their employment in labour markets outside the village. If many households are simultaneously trying to increase their off-farm labour allocations in villages affected by the NFPP, it is likely that within these individual villages, either wages would diminish

Table 11.5 Impact of NFPP on Off-Farm Employment

	DID no covariates	DID with covariates	PS matching, kernel regression	PS matching, local linear regression	PS weighting
Number of HH members in off-farm employment	0.130 (0.150)	0.235 (0.164)	0.0907 (.166)	0.137 (0.151)	0.177 (0.196)
Total days off-farm employment per HH	85.30 (47.14)*	101.83 (51.08)**	55.09 (50.65)	69.02 (52.36)	71.13 (59.32)
Number of HH members working outside village	0.281 (0.142)**	0.369 (0.157)**	0.247 (0.162)	0.251 (0.156)	0.283 (0.178)
Total days worked outside village per HH	105.15 (45.55)**	129.88 (49.18)***	83.10 (47.72)*	84.23 (49.38)*	85.77 (55.90)

* significant at 10%
** significant at 5%
*** significant at 1%

rapidly or jobs would be rationed. In contrast, the regional or national labour market, which is mainly located in urban areas, would be much less affected by the increase in labour supply resulting from the ban on logging.

11.5.2 Relationship Between Employment Impacts and Income Impacts

So far in this section we have discussed why the off-farm labour allocation of households might be expected to increase in response to the ban on harvesting timber, and the empirical results have shown that an increase in off-farm employment outside the home village has occurred in areas affected by the NFPP. However, we would not expect an increase in income to occur as a result of the shift of labour from forest activities to off-farm employment. With sufficient demand for off-farm labour, total income may remain constant if household members who are no longer required for timber harvesting are re-employed elsewhere at a wage that is equal to the previous return to their labour in forest activities. Nevertheless, it does not provide an explanation for the observed increase in total income. This is because if the off-farm wage rate was higher than the return to labour in forest activities before the logging ban, we would expect households to have already allocated more labour to off-farm employment. As the NFPP does not alleviate any constraints in relation to labour market access, then if household income could be increased by moving from forest activities to off-farm employment, such a change in labour allocation should be observed in both NFPP and non-NFPP households.

An explanation for the increase in income resulting from the NFPP is that households are not simply comparing the returns to labour in alternative occupations.

Instead, there are a range of further factors that are accounted for in the decision making process.

The first of these factors is the potential cost involved in seeking off-farm employment. This is likely to be particularly important in relation to employment outside the village. Rather than directly comparing earnings from farm and off-farm activities, households will compare expected income net of expenditure. The income data used in this chapter are net of expenditure on inputs to agriculture and forest activities, but do not include the costs of seeking off-farm employment. Where individuals look for work outside the village, these costs may include travel expenses and a higher cost of living at the destination. Because the income data do not account for these costs, the empirical finding of increased income in NFPP areas may disguise a reduction in consumption by households because a greater proportion of their total income is spent on enabling household members to live and work outside the village.

Another category of additional factors consists of non-monetary reasons why household members prefer to work in forest or agricultural activities on household land rather than seek off-farm employment. These reasons may include preferring to work for themselves rather than someone else, or perceived riskiness of off-farm income, as employment may not be secure. As with the financial costs of seeking employment, non-monetary factors are likely to be more significant for employment outside the village. The standard Harris–Todaro model of migration (Harris and Todaro, 1970) involves the comparison of the expected wage at the destination with the return to labour at home. However, there have been many related models that emphasise additional variables. In fact, Todaro (1969) stresses that real incomes, taking account for differences in the costs of living, should be compared, rather than nominal incomes. He also acknowledges that factors such as relative living conditions will affect the migration decision as well as earnings differences.

More specifically, Sjastaad (1962) describes how the 'psychic costs' of migrating mean that there will be a minimum level of earnings at the home location that will leave an individual indifferent between migrating and not migrating. If home earnings fall below that level, the earnings differential will persuade the individual to migrate. 'Psychic costs' are the negative welfare effects of leaving family and friends, and they imply earnings differentials that are larger than the monetary costs and benefits of migration would suggest. Importantly, such earnings differentials do not suggest misallocation of resources because they are the result of the preferences of individuals or households. Mundlak (1979) analyses migration based on comparisons of expected income, as in the Harris–Todaro model. However, he also includes a measure of 'quality of life' in the model used. 'Quality of life' may be assumed to be higher in rural or in urban areas. In the former case, a premium is required in addition to urban (off-farm) wage rates to encourage migration, while in the latter case, urban wage rates may be lower.

If non-monetary factors such as differences in quality of life or 'psychic costs', in combination with earnings differentials, affect the decision to migrate we would expect to see an increase in off-farm employment in response to a fall in income from forest activities, as discussed previously. However, the increase in off-farm employment could in this case lead to an increase in the total income of the household. In this context, the rise in total income would not necessarily indicate an increase in the overall welfare or utility of the household because it may be accompanied by a reduction in non-monetary benefits or an increase in non-monetary costs associated with migration. Similarly, if migrating to earn the higher income involves expenditure on travel costs and living costs, with a resulting reduction in consumption, welfare may not rise along with income.

11.6 Conclusion

The initial driver behind this chapter was the recognition that while much effort frequently goes into valuation of the benefits of forest conservation policies, the negative impacts or costs for individual households living in affected forest areas are often not accounted for. Furthermore, existing studies and reports on the NFPP suggested that the ban on logging was having serious detrimental impacts on the livelihoods of such households. We therefore made use of programme evaluation methods to identify the impacts of the NFPP on two interrelated facets of household livelihoods, namely income and off farm labour supply decisions.

Overall, our results show that the NFPP has had a significant, although fairly small, negative impact on the income earned from timber harvesting by households in collective forest areas. More surprisingly, it has had a positive impact on total household incomes in villages affected by the NFPP. These are potentially very important findings because the belief that the NFPP has been damaging for forest-based households, and has increased levels of rural poverty by denying access to a key source of income, has led to calls for its removal, despite evidence of environmental benefits. If the impacts of the programme are in fact less harmful than previously thought, then the case for its removal is significantly weakened. However, the differences between the results of this study and the conclusions of other studies raise two questions. The first of these is why the impacts of the NFPP on timber incomes are small. The second question is why the loss of timber incomes has not resulted in a reduction in overall incomes, and has in fact led to an overall increase.

The main explanation for the small change in incomes from timber resulting from the NFPP is that prior to the ban on logging, timber incomes were already low in both NFPP and non-NFPP areas. There are two reasons for this: firstly, previously high rates of deforestation meant that most households only have access to poor quality, inaccessible or immature timber. The rapid deforestation in the Southern collective forest provinces has been documented by Rozelle et al. (2000), and the responses to our household survey also indicated that much of the timber on the respondents' forest land had already been harvested. The second reason for the low share of timber in total incomes was that structural changes were already occurring in the rural economy, with the result that increasing proportions of household incomes were coming from off-farm employment. This has meant that although increases in income from employment have been greater in NFPP areas, timber also contributed to a smaller share of total household income by 2004 in non-NFPP areas.

The impact of previous deforestation on initial timber incomes may also provide a partial explanation for the difference between our findings and those of previous studies. Guizhou Province, where our study took place, is fairly representative of the southern collective forest provinces in China, in terms of forest management and past harvesting patterns. However, many of the other studies were carried out in the south-western provinces of Sichuan and Yunnan, which have historically had different patterns of forest management and land use. In particular, deforestation and forest degradation have been more severe in the southern provinces. In the south-western provinces, forest cover remains higher therefore the potential for households to earn income from timber activities, and correspondingly the potential losses from a ban on those activities, is greater. The findings of this study therefore can to some extent be extrapolated to other southern forest provinces, but the impacts may be different in the southwest provinces. The impacts in the forest region of northeast China are expected to be yet more different because the forests are state, rather than collective, owned and managed and because alternative employment opportunities are much more limited.

The other difference between this study and previous studies is that total incomes are found to increase as a result of the NFPP instead of fall. This is contrary to the qualitative studies cited in Section 11.2 and also to the views held, for example by county level forestry officials who were interviewed as part of this project. One factor is likely to be that those interviewed in a case study setting about what they believe to be the impacts of the programme assume that there must have been income losses because households are no longer able to undertake income generating activities that they were previously allowed. However, this does not take account of the alternative uses for the labour that was previously used for forest activities. An important implication of this is that in order to understand the equilibrium effects of a programme on household incomes as opposed to simply the direct effects, it is necessary to collect data on total sources of income as the final impacts may be different to the timber income impacts in isolation.

Lastly, in this chapter we considered how labour that is no longer required for timber management and harvesting is reallocated into alternative activities. We find that the reduction in forest activities has been associated with a rise in off-farm employment, specifically outside the village. The standard model of labour allocation would predict the effect on employment, but would not suggest an increase in overall household income. However, where households make labour allocation decisions, in particular those relating to migration out of the village, factors other than relative earnings in alternative activities will be taken into consideration. Households are assumed to compare all monetary and non-monetary costs and benefits of migrating versus not migrating. If expenditure on travelling to sources of off-farm employment, higher costs of living in the destination locations, or poorer living conditions away from the village are important, household members may not migrate even with the potential for higher earnings. However, a drop in forest-related income would lead to migration by members of marginal households, and in this context, would result in an increase in overall income. Despite this increase in income, we cannot assume that the overall welfare of the household has been increased.

There are a number of final caveats to bear in mind. The first of these is that the results in this chapter represent average impacts across households. Some households are likely to have been affected to a greater extent than suggested, because they have previously specialised on timber production and have lost investments, or because their alternative income generation opportunities are more limited. The first of these may have implications for the long term environmental impacts of the programme. If the ban creates disincentives to invest in timber plantations then forest cover will not increase over time without the continuous involvement of the state. The second factor is important in relation to poverty alleviation in rural areas such as those where the NFPP has been implemented.

Another caveat is that these results pertain to a specific province in China. We have argued that they may be broadly representative of the southern collective forest areas. However, the different socio-economic, institutional and environmental conditions in the southwest and northeast forest areas of China mean that the effects of the ban may have been different.

A final issue is that even if the ban on logging does not reduce household incomes overall, it can be argued to infringe on the rights that the households hold to forest land. Through the Household Responsibility System, they were allocated the rights to harvest timber on their plots of forest land, and in many cases provided contracts for 30 years or more. That land use rights have been removed without compensation may have implications in terms of equity or in terms of incentives to sustainable manage forest or other types of land in future.

Acknowledgments   We would like to acknowledge the financial support of the China Council for International Cooperation on Environment and Development as well as the assistance of the School of Environmental Sciences of Beijing University in implementing the survey used in this chapter and particularly the help received by Ms. Tao Wendi. We are also grateful for the comments received by Prof. Jeff Bennett, Prof. Erwin Bulte, Dr. Ben Groom, Dr. Shinwei Ng, Prof. David Pearce, Prof. Jerry Warford, and Prof. Jintao Xu. We are also grateful for the comments received from the participants of the International Symposium on Evaluating China's Ecological Restoration Programs (hosted by the Forest Economics and Development Research Center of China's State Forestry Administration and Michigan State University in Beijing Oct 18–20) and in particular Prof. Runsheng Yin, Dr. Emi Uchida, Prof. David Newman and Prof. Bill Hyde.

References

Abadie, A. (2005). Semiparametric difference-in-differences estimators. *Review of Economic Studies, 72,* 1–19.

Ashenfelter, O. (1978). Estimating the effect of training programs on earnings. *Review of Economics and Statistics, 60*(1), 47–57.

Ashenfelter, O., & Card, D. (1985). Using the longitudinal structure of earnings to estimate the effect of training programs. *Review of Economics and Statistics, 67,* 648–660.

Becker, S. O., & Ichino, A. (2002). Estimation of average treatment effects based on propensity scores. *The Stata Journal, 2*(4), 358–377.

Blundell, R., & Costa Dias, M. (2000). Evaluation methods for non-experimental data. *Fiscal Studies*, *21*(4), 427–468.

Caliendo, M. and S. Kopeinig (2008). "Some Practical Guidance for the Implementation of Propensity Score Matching." Journal of Economic Surveys 22(1): 31–72.

Chen, L., Xiang, C., Liu, X. & Mu, K., 2001. *Case Study on the Natural Forest Protection Program in Sichuan Province's Li County, Pingwu County and Chuanxi Forestry Bureau*, CCICED Western China Forest and Grasslands Taskforce.

Dehejia, R. H. and S. Wahba (2002). "Propensity Score-Matching Methods for Nonexperimental Causal Studies." Review of Economics and Statistics *84*(1): 151–161.

Demurger, S., & Fournier, M. (2003, August 28–29). *Forest protection and rural economic welfare: The case of Labagoumen nature reserve (China)*. Fourth Bioecon Workshop on the Economics of Biodiversity Conservation, Economic Analysis of Policies for Biodiversity Conservation, Venice, Italy.

Ferraro, P. J. (2002). The local costs of establishing protected areas in low-income nations: Ranomafana National Park, Madagascar. *Ecological Economics, 43*(2–3), 261–275.

Harris, J.R. & Todaro, M.P., 1970. Migration, Unemployment and Development: A Two-Sector Analysis. *American Economic Review, 60*(1), 126–142.

Heckman, J., Ichimura, H., & Todd, P. (1997). Matching as an econometric evaluation estimator: Evidence from evaluating a job training programme. *Review of Economic Studies, 64*, 605–654.

Heckman, J. J., & Robb, R. (1984). Alternative methods for evaluating the impact of interventions. In J. Heckman & B. Singer (Eds.), *Longitudinal analysis of labor market data.* (pp. 156–245). Cambridge: Cambridge University Press.

Hirano, K., G. W. Imbens and G. Ridder (2003). "Efficient Estimation of Average Treatment Effects Using the Estimated Propensity Score." Econometrica *71*(4): 1161–1189.

Innes, R., Polasky, S., & Tschirhart J. (1998). Takings, compensation and endangered species protection on private lands. *Journal of Economic Perspectives, 12*(3), 35–52.

Katsigris, E. (2002). Local-level socioeconomic impacts of the Natural Forest Protection Program. In J. Xu, E. Katsigris, & T. A. White (Eds.) *Implementing the natural forest protection program and the sloping lands conversion program.* Western China Forests and Grassland Taskforce of China Council for International Cooperation on Environment and Development, Beijing.

Katsigris, E., Bull, G. Q., White, A., Barr, C., Barney, K., Bun, Y., Kahrl, F., King, T., Lankin, A. & Lebedev, A. (2004). The China forest products trade: Overview of Asia-Pacific supplying countries, impacts and implications. *The International Forestry Review, 6*(Part 3/4), 237–253.

Kramer, R. (1996). *Slowing tropical forest biodiversity losses: Cost and compensation considerations*. Workshop on Biodiversity Loss.

Miao, G. & West, R. A. (2004). Chinese collective forestlands: contributions and constraints. *International Forestry Review, 6*(3/4), 282–298.

Mundlak, Y. (1979). *Intersectoral factor mobility and agricultural growth*. Washington, DC: International Food Policy Research Institute.

Quandt, R. E. (1972). A new approach to estimating switching regressions. *Journal of the American Statistical Association, 67*, 306–310.

Rosenbaum, P. R., & Rubin, D. B. (1983). The central role of the propensity score in observational studies for causal effects. *Biometrika, 70*, 41–55.

Roy, A. D. (1951). Some thoughts on the distribution of earnings. *Oxford Economic Papers*, 3, 135–146.

Rozelle, S., Huang, J., Husain, S. A., & Zazueta, A. (2000). *China: From afforestation to poverty alleviation and natural forest management.* Washington, DC: World Bank.

Rubin, D. B. (1978). Bayesian inference for causal effects: The role of randomization. *The Annals of Statistics, 6*(1), 34–58.

Shen, M. (2001). *How the logging ban affects community forest management: Aba prefecture, North Sichuan, China*. Policy frameworks for enabling successful community-based resource management initiatives, East-West Center.

Singh, I., Squire, L., & Strauss, J. (1986). *Agricultural household models: Extensions, applications, and policy.* Baltimore: Johns Hopkins University Press.

Sjastaad, L. A. (1962). The cost and returns of human migration. *The Journal of Political Economy, 70*(5), 80.

Swanson, T., Zhang, S., Kontoleon, A., Mullan, K. & Ng, S. W., 2006. *Assessing the impacts of the Natural Forest Protection Programme on local community welfare,* CCICED Task Force on Environmental and Natural Resources Pricing and Taxation.

Todaro, M.P., 1969. A Model of Labor Migration and Urban Unemployment in Less Developed Countries. *American Economic Review, 59*(1), 138–148.

Wang, X., Han, H. & Bennett, J., 2008. China's Land Use Management. In J. Bennett, X. Wang, & L. Zhang, eds. *Environmental Protection in China: Land Use Management.* Cheltenham, UK; Northampton, MA: Edward Elgar.

Weyerhaeuser, H., Wen, S. & Kahrl, F. (2006). *Emerging forest associations in Yunnan, China: Implications for livelihoods and sustainability.* Kunming, Yunnan: International Institute for Environment Development.

Xu, J., Katsigris, E., & White, T. A., Eds. (2002) *Implementing the natural forest protection program and the sloping lands conversion program.* Western China Forests and Grassland Taskforce of China Council for International Cooperation on Environment and Development, Beijing.

Yang, Y. (2001). Impacts and effectiveness of logging bans in natural forests: People's Republic of China. In P. B. Durst, T. R. Waggener, T. Enters, & T. L. Cheng (Eds.) *Forests out of bounds: Impacts and effectiveness of logging bans in natural forests in Asia-Pacific.* Bangkok: FAO Asia-Pacific Forestry Commission.

Zhang, P., Shao, G., Zhao, G., Le Master, D., Parker, G., Dunning, J. & Li, Q. (2000). China's forest policy for the 21st century. *Science,* 288(5474), 2135–2136.

Zuo, T. (2002). Implementation of the NFPP. In J. Xu, E. Katsigris, & T. A. White, Eds. (2002) *Implementing the natural forest protection program and the sloping lands conversion program.* Western China Forests and Grassland Taskforce of China Council for International Cooperation on Environment and Development, Beijing.

Chapter 12
An Estimation of the Effects of China's Forestry Programs on Farmers' Income

Can Liu, Jinzhi Lü, and Runsheng Yin

Abstract In the late 1990s, the Chinese government initiated some new programs and consolidated other existing programs of ecological restoration and resource development in its forest sector, and named them as "Priority Forestry Programs," or PFPs. They include the Natural Forest Protection Program (NFPP), the Sloping Land Conversion Program (SLCP), the Desertification Combating Program around Beijing and Tianjin (DCBT), and the Wildlife Conservation and the Nature Reserve Development Program (WCNR). In addition to improving the environmental and resource conditions, a frequently reiterated goal of these PFPs is to enhance the income of rural residents. Thus, a question of great interest is: How has implementing the PFPs affected the farmers' income and poverty status? The objective of this chapter is to address this question, using a fixed-effects model and panel data from over 2,100 households in ten counties of Sichuan, Hebei, Shaanxi, and Jiangxi Provinces. The empirical evidence indicates that their effects are mixed. The impacts of the SLCP, the NFPP, and the DCBT are significantly positive, whereas the impact of the WCNR is negative and the SBDP has little effect on household income. Furthermore, these impacts show substantial variations in different counties. Also, land for home gardening, labor for off-farm employment, and technical and institutional changes play major roles, implying that more attention should be directed to increasing income from cash crops, off-farm employment, and training and extension.

Keywords Ecological restoration · Priority Forestry Programs · Farmers' incomes · Poverty alleviation · Panel data · Fixed-effect model

C. Liu (✉)
China National Forestry Economics and Development Research Center,
State Forestry Administration, Beijing, P.R. China
e-mail: liucan@public.bta.net.cn

R. Yin (ed.), *An Integrated Assessment of China's Ecological Restoration Programs,*
DOI 10.1007/978-90-481-2655-2_12, © Springer Science+Business Media B.V. 2009

12.1 Introduction

China has achieved spectacular economic growth since it initiated economic reforms in 1978. The total gross domestic product (GDP) in 2000 grew to more than seven times that in 1979; and it further increased by 74.5% from 2000 to 2006, when it reached almost 21 trillion yuan. Similarly, farmers' annual per capital income increased from 134 yuan in 1978 to 3,587 yuan in 2006 (China National Statistics Bureau, or CNSB, 2007). Based on the national standard, China's population living in absolute poverty was over 250 million in 1978, but was reduced to 21.5 million by the end of 2006 (CNSB, 2007).

Despite these tremendous successes, poverty, especially in the rural areas of western China, remains a troublesome problem. According to the World Bank, rural residents in China living below its poverty line of $1 per day still amounted to 135 million in 2004 (Chen & Ravillion, 2004). The "Human Development Report 2005" of the United Nations Development Program (UNDP) noted that the pace of poverty alleviation in China has slowed down markedly over the last decade (UNDP, 2005). Obviously, how to further reduce rural poverty and increase farmers' income is still a top priority in China.

Often, remote and mountainous areas are connected with both abundant forests and acute poverty (FAO, 2006; World Bank, 2000). In China, most of its 592 poverty countries are found in these areas (State Forestry Administration, or SFA, 2003). Indeed, in many impoverished areas, forestry is the main source of income for farmers (Liu & Lü, 2008). In the late 1990s, the Chinese government initiated some new programs and consolidated other existing programs of ecological restoration and resource development in its forest sector, and named them as "Priority Forestry Programs," or PFPs (SFA, 2002). In addition to improving the environmental and resource conditions, a frequently reiterated goal of these PFPs is to enhance the income of rural residents (SFA, 2002). Thus, a question of great interest is: How has implementing the PFPs affected the farmers' income and poverty status? The objective of this chapter is to address this question empirically.

Natural disasters in the late 1990s intensified an environmental debate in China and triggered the government to initiate the Natural Forest Protection Program (NFPP) in 1998 and the Sloping Land Conversion Program in 1999 (Yin, Xu, Li, & Liu, 2005). Following successful piloting during 1998–1999, the NFPP was formally launched in 2000, with an initial investment of 96.4 billion yuan for the decade (Yin & Yin, 2008).[1] A key component of the NFPP is logging bans over 30 million ha of natural forests in the upper reaches of the Yangtze River and the upper/middle reaches of the Yellow River. In other areas, harvest restrictions are imposed. The SLCP was piloted in Sichuan, Shaanxi, and Gansu provinces in 1999.[2] The primary goal of the program is to convert 14.6 million ha of sloping and desertified farmland

---

[1] This is equivalent to roughly US$14.1 billion given the current exchange of $1 = 6.85 yuan.

[2] Also known as the "Grain for Green" program in the international literature.

into forest and grass coverage from 2001 to 2010. When it was formally launched in 2002, the SLCP was expanded to 25 provinces, with an original budget of 225 billion yuan (Yin & Yin, 2008).

In addition to the above two mega-programs, a number of other efforts of ecological restoration and forest expansion have been consolidated into the following four programs: the Desertification Combating Program around Beijing and Tianjin (DCBT), the Shelterbelt Development Program in the Three-Norths[3] and the Yangtze River Basin (SBDP), the Wildlife Conservation and Nature Reserve Program (WCNR), and the Industrial Timber Plantation Program (ITPP). Together with the NFPP and the SLCP, these programs comprise the six PFPs (SFA, 2004, 2005), which are implemented for the purpose of improving the environmental conditions, increasing farmers' income, and boosting domestic timber supply. The different policy arrangements in implementing these six PFPs are summarized in Table 12.1.

Table 12.1 Key Policy Measures of the PFPs

Program	Key policies
Sloping Land Conversion Program (SLCP), covering 25 provinces during 2001–2010	• Sloping or desertified cropland is converted into ecological and/or economic forest, and grassland; ecological forest should account for 80% of total converted land. • The central government subsidizes farmers in the form of seeds or seedlings, grain, and cash. • Subsidies last 8 years for ecological forest, 5 years for economic forest, and 2 years for grassland. The annual cash subsidy is 300 yuan/ha, and the annual grain subsidy is 1,500 kg/ha in the Yellow River basin and 2,250 kg in the Yangtze River basin. • The central government also makes fiscal transfers to compensate the entailed losses to local fiscal revenues. • Estimated total investment is 225 billion (US$32.8 billion).
Natural Forest Protection Program (NFPP), covering 17 provinces during 2000–2010	• Complete ban on commercial logging in the upper Yangtze and middle Yellow River basins and sharp reduction in commercial harvests in other program areas. • Shutting down of certain processing facilities, compensating logging firms, and disposing displaced workers and equipment. • Promotion of afforestation and forest management wherever possible. • Strengthening administration and law enforcement, including forest protection. • Restricting the forest industry, and improving the efficiency of timber utilization. • Initial investment commitment is 96.4 billion (US$14.1 billion).

[3] The "Three Norths" are the northwestern, north-central, and northeastern regions of China.

Table 12.1 (continued)

Program	Key policies
Wildlife Conservation & Nature Reserve Development Program (WCNR), scattered all over the country during 2001–2050	• Priority protected areas are administrated by the central government, while smaller and less critical areas are managed by the local governments. • Established reserves will reach 1,800 by 2010, 2,000 by 2030, and 2,500 by 2050. • Wetland protection and restoration, ecotourism development, and wildlife breeding. • Encouraging domestic and international participation and contributions, including broad involvement of the private sector. • Balancing ecosystem conservation with socioeconomic development. • Strengthening the role of science and technology, particularly nature reserve and biodiversity monitoring and evaluation. • Total planned investment is 135.65 billion yuan (US$19.8 billion), with the central government covering 66.44 billion yuan.
Shelterbelt Development Program (SBDP), covering all 31 provinces during 2001–2010	• Including shelterbelt programs in the Three Norths (northwest, north, and northeast), the Yangtze River basin, the Zhujiang River basin, and the Taihang Mountain Range. • Mobilization of public agencies, civil society, individuals to contribute to the shelterbelt development and tree planting. • Encouraging local government investment and local labor contribution, and adopting new silvicultural techniques. • Total planned investment is 70 billion yuan (US$10.2 billion).
Desertification Combating around Beijing and Tianjing (DCBT), including Inner Mongolia, Hebei, Shanxi, Beijing, and Tianjin during 2001–2010	• Converting desertified land into forestland and grassland by means of flexible and diversified measures based on the local conditions. • Changing herding and animal husbandry practices to control overgrazing and rehabilitate degraded grassland. • Developing irrigation projects, and resettling people away from fragile areas. • Extension of suitable production technology and energy sources. • Establishing desertification monitoring and dust storm forecasting systems. • Total projected investment is 57.7 billion yuan (US$8.4 billion).
Industrial Timber Plantation Development Program (ITPP), covering 18 provinces during 2001–2015	• Market-driven and profit-orientated efforts for increasing domestic timber supply. • As high as 70% of the investment may come from subsidized National Development Bank loans, with 20% from direct government funding and 10% from other sources; in addition, tax incentive is provided. • Encouraging active participation by various enterprises – state or collectively owned, shareholder based, or fully private. • Planned area of establishment is 4.69 million ha by 2005, 9.2 million ha by 2010, and 13.33 million ha by 2015. • Projected total investment is 71.8 billion (US$10.5 billion).

Several scholars have investigated the impact of the SLCP on farmers' income and livelihoods. In addition to examining its cost effectiveness and sustainability, Uchida, Xu, & Rozelle (2005) and Uchida, Xu, Xu, and Rozelle (2007) analyzed its influence on eradicating poverty in the countryside. They found that the program has been successful in poverty alleviation, even though poor households may not have benefited the most. Further, their evidence shows that rural households participating in the program have already begun transferring their labor to non-farming sectors more rapidly than those not participating in the program. In contrast, using data collected from Sichuan, Shaanxi, and Gansu for the first few years of the program, Xu, Tao, & Xu, (2004) found that it made little difference in affecting farmers' income between non-participating and participating households. Their conclusion thus suggests that the SLCP's role in relieving poverty is limited; the reduction of poverty is more likely driven by the overall economic development, which provides greater opportunities to farmers, rather than the direct subsidies of the land conversion.

Observers also point out that in some cases, the goal of the program is not well understood by farmers; and it may even be inconsistent with their aspirations, which have affected their enthusiasm for participation (Du, 2004). According to Xu (2003), the main reasons that some farmers lack interest in the program are partly because the subsidies are not delivered on time and in full, and partly because no appropriate remedies were put in place to address the restrictions on intercropping in the forested land and gathering fuelwood. These factors have led to an adverse effect on the livelihoods of the farmers, who not only rely heavily on forest resources but also tend to be among the poorest rural population (Hu, 2002). Based on the case studies of Nujing and Baoshan in Yunnan, Weyerhaeuser, Wilkes, & Kahrl (2005) also revealed negative impacts of the SLCP and the NFPP on the livelihoods of highland communities.

On the other hand, the study by Zhi & Shao, (2004) claimed that the income of farm households would be significantly improved during the time when they receive the government compensation. If the subsidies are terminated after the program expires in 5–8 years and farmers are not allowed to utilize their retired lands for economic purposes, they could suffer a loss. In the authors' opinion, the policy that mandates that the proportion of economic forest be no more than 20% has failed to consider the regional disparity and the basic fact that the country has a large rural population but a relatively small amount of cropland. Thus, the government must take steps to improve farmland quality and increase farming productivity in order to address the problem of food supply following the land conversion. Li (2001) further proposed that the SLCP should be made up of two explicit, complementary parts – a restoration component pursuing ecological benefits and a production component pursuing economic benefits. He also argued that the short time of compensation is a constraint to livelihood improvement, which is in sharp contrast with the 30-year compensation of a similar conservation program in the United Kingdom.

Compared to the SLCP, there have been fewer studies of the socioeconomic impacts of the NFPP and other programs. Using household data, Liu et al. (2005) and Ni, Wang, & Yang (2002) found that the NFPP had a negative effect on the income of farmers living close to the protected natural forests. With data from 18 counties in Shanxi, Inner Mongolia, and Hebei, Liu and Zhang (2006) showed that the DCBT had a positive effect on farmers' income, but the effect varied from county to county, due in part to uneven program investments. However, they did not consider all of the main factors affecting farmers' income, including production inputs and household characteristics. In addition, little work has been conducted so far to make a comparative analysis of the impacts of the PFPs. In fact, most of the existing studies have focused on a single PFP, with few studies dealing with two. This is unfortunate given the fact that these PFPs have all somehow affected farmers' income and their impacts may be interlinked and variable. While some programs have positive income impacts, others can cause negative or insignificant effects. So, if we consider the impact of only one or two programs, we may derive biased results and thus misunderstand the impacts of different PFPs.

Here, we use a fixed-effect model to estimate the income impacts of the six PFPs, with a panel dataset of over 2,100 households in ten counties of four provinces. Broader cross sections and longer time series are two unique features of the dataset. That is, every sample county has at least two PFPs and covers a period of 10 years from 1995, before the PFPs were initiated or consolidated, to 2004, when their implementation was well underway. By removing the influence of control variables, including production inputs and household and village characteristics, based on such a large dataset, our model enables us to identify the specific effect of each program on farmers' income in an unbiased and robust way. Our results indicate that different programs have quite different impacts. While the WCNR has a negative impact on household income and the effect of SBDP is insignificant, other PFPs have made positive contributions to household income, among which the SLCP has the largest impact on household income, followed by the DCBT. The NFPP also has a positive effect on farmers' income, contrary to what has been found previously. The chapter is organized as follows: The next section is devoted to method and data, the third section reports our empirical results, and we present our conclusions and discussion in the final section.

12.2 Method and Data

Generally speaking, farmers' income is determined by their production inputs and other biophysical and socioeconomic factors. Production inputs constitute labor, capital, and land. Included in land are farmland, forestland, and other land mainly for growing vegetables and fruits. In addition, land-based production activities entail cash outlays for commercial seeds, fertilizers, plastic sheets, and the like. Moreover, household and village characteristics affect farmers' income. For example, as part of the human capital, educational attainment is an important household feature (Schultz, 1964). And biophysical and institutional variables at the village level, like plot size of farmland,[4] topography, and road condition, are also relevant to income determination. Of course, implementing the six PFPs have affected farmers' income (see Table 12.1), even though the direction and magnitude of each program's impact may differ.

To summarize, it can be stated that farmers' income (R) is principally a function of production inputs $(X_1, \ldots, X_j, \ldots, X_J)$, household and village characteristics $(Y_1, \ldots, Y_k \ldots, Y_K)$, and engagements in the PFPs $(Z_1, \ldots, Z_m, \ldots, Z_M)$. So, our conceptual model is:

$$R_{it} = e^{\alpha_0} \prod_{j=1}^{J} X_{it}^{\alpha_j} \prod_{k=1}^{K} Y_k^{\beta_k} \prod_{m=1}^{M} Z_{mt}^{\gamma_m} \varphi_{it} \qquad (12.1)$$

where i is the ith sample household $(i = 1, 2, \ldots I)$, t is the time period $(t = 1, 2, \ldots, T)$, α_j, β_k, and γ_m are coefficients to be estimated, and ϕ_{it} is the error term that is assumed to be independent and identically distributed.

In accordance with the distributions of farmers' incomes and the six PFPs, we selected the following ten counties in Jiangxi, Hebei, Sichuan, and Shaanxi provinces for our surveys, using a stratified random sampling method: Zhangbei and Pingquan in Hebei; Xiushui, Xingguo, and Suichuan in Jiangxi; Zhen'an and Yanchang in Shaanxi; and Nanbu, Nanjiang, and Mabian in Sichuan (see Table 12.2). These are national poverty counties, where agriculture still is a major source of income, accounting for more than one-third of their GDP. Implementing the PFPs, especially the SLCP, has caused a sharp farmland reduction. For instance, the share of farmland in the sample counties of Sichuan has declined from 39.6% in 1998 to 26.0% in 2004 and the share of farmland in the sample counties of Shaanxi has declined from 37.3% to 21.5% during the same period. In turn, this has resulted

Table 12.2 Participation of Sample Counties in the PFPs

Province	county	NFPP	SLCP	DCBT	WCNR	ITPP	SBDP
Hebei	Zhangbei	×	✓	✓	×	×	✓
	Pingquan	×	✓	✓	×	×	✓
Jiangxi	Xiushui	×	✓	×	×	✓	✓
	Suichuan	×	✓	×	×	✓	✓
	Xingguo	×	✓	×	×	✓	✓
Shaanxi	Yanchang	✓	✓	×	×	×	✓
	Zhenan	✓	✓	×	✓	×	✓
Sichuan	Mabian	✓	✓	×	✓	×	✓
	Nanbu	✓	✓	×	×	×	✓
	Nanjiang	✓	✓	×	×	×	✓

Note: ✓ indicates that the sample county participates in the PFP; × indicates otherwise.

[4]In implementing the household responsibility system in the late 1970s and early 1980s, farmland and forestland were divided into small plots and then allocated to individual families.

in a reduction of food production and income generated from farming. Except for Zhangbei, where three townships were selected, six townships were chosen in each county. We conducted surveys of 2,700 households in 171 villages of 57 townships. The data cover 10 years from 1995 to 2004.

After removing those with incomplete information, we had over 2,100 valid household surveys for this study – the specific number of valid surveys differed from year to year (see Table 12.3). Over the whole sampling period, more and more households were involved in the PFPs as these programs were implemented; and, while some participated in multiple PFPs, others did not participate in any of them. Table 12.3 also indicates that a large number of households took part in the NFPP and the SLCP, but only a few took part in the SBDP and none in the ITPP. It seems odd that only a few sample households were involved in the SBDP, which is widespread. This is due in large part to the fact that protective shelterbelts are mostly established on public lands, instead of lands devolved to households. Similarly, as a consequence of the limited extent of the ITPP in combination with our random drawing, the ITPP was not captured by any sample households. Thus, it will be excluded from the following empirical estimation.

Included in the data are total household income; farmland area, forestland area, and area of other land mainly for home gardening; labor inputs, including land-based labor use, labor use for other local production activities, and labor for off-farm employment; cash outlays for land-based activities, such as expenses for commercial seeds, fertilizers, plastic sheets, and pesticides; and PFPs' activity

Table 12.3 Distribution of Valid Sample Households by PFPs

Year	NFPP	SLCP Overall	Yellow River basin	Yangtze River basin	DCBT	WCNR	SBDP	Total
1995						78		2,137
1996						78		2,140
1997		8		8		78	1	2,151
1998	615	10		10		80	1	2,173
1999	619	364	136	228		82	1	2,212
2000	1,032	536	186	350		87	1	2,252
2001	1,033	597	186	411	1	87	1	2,268
2002	1,045	782	217	565	126	87	8	2270
2003	1,043	985	235	750	225	88	11	2,223
2004	1,046	951	203	748	236	88	9	2,232

Notes:
1. NFPP, SLCP, DCBT, WCNR, and SBDP represent, respectively, the Natural Forest Protection Program, the Sloping Land Conversion Program, the Desertification Combating Program around Beijing and Tianjin, the Wildlife Conservation and Nature Reserve Program, and the Shelterbelt Development Program in the Three-Norths and the Yangtze River Basin.
2. The Industrial Timber Plantation Program was not included because none of the sample households participated in it. While the SBDP is widely distributed, it was not much captured at the household level given the nature of the project.

variables – effective area enrolled in each program, except for the WNRP, for which the relevant variable is defined as the inverse of the distance to the nearby nature reserve. Information on family characteristics, such as household size, education of household head, and status as a village leader or party member, was also gathered. Similarly, village-level features, such as topography (i.e., plain, hill, or mountain), distance to the nearest township, and road condition, were captured. Additionally, we obtained information regarding the average size of forestland and farmland per plot of the sample households, which can partially reflect the production efficiency and economies of scale of agriculture and forestry. Total incomes and cash outlays of sample households are deflated and converted to the 1994 constant yuan, using the rural consumer price index and rural industrial product price index from the Chinese Statistical Yearbooks published by the China National Statistical Bureau (http://www.stats.gov.cn).

Preliminary statistics of the household data for 1995 and 2004 are summarized in Table 12.4. Due to sample differences in biophysical and socioeconomic conditions, the data have large variances. However, our data description below focuses on their mean value changes over time. The average household income grew from 4,291.1 yuan in 1995 to 7,569.2 yuan in 2004. The mean farmland reduced from 9.5 mu in 1995 to 5.5 mu in 2004; in the meantime, the average forestland increased from 10.5 mu to 15.2 mu, and the mean of other land increased from 0.4 to 2.5 mu. [5] Clearly, much of the lost farmland has been converted into forestland or other land. The mean area of farmland per plot declined from 1.6 mu in 1995 to 1.4 mu in 2004, while the mean area of forestland per plot increased from 3.7 to 4.4 mu during the same period. Labor input for land-based activities declined slightly, and it more than doubled for off-farm activities. Expenditure for fertilizers, seeds, and other cash inputs increased significantly as well. In 2004, the average area enrolled in the SLCP in the Yellow River basin was 7.2 mu, while the figure in Yangtze River basin was 2.1 mu. This indicates that farmers in the former region have benefitted more than those in the latter, even though grain subsidy in the latter is a bit higher (see Table 12.1). The density distributions of household income for 1996 and 2001, which are nearly normal, are given in Fig. 12.1.

Following log transformation and adding a time trend variable to capture the effect of technical and institutional changes on household income, our empirical model is specified as:

$$\ln R_{it} = \alpha_0 + \alpha_1 \ln X_{1it} + \alpha_2 \ln X_{2it} + \cdots + \alpha_7 \ln X_{7it} + \beta_1 \ln Y_{1it} + \beta_2 \ln Y_{2it} + \cdots$$
$$+ \beta_5 \ln Y_{5it} + \gamma_1 \ln Z_{1it} + \gamma_2 \ln Z_{2it} + \cdots + \gamma_{10} \ln Z_{10it} + \theta T + \varphi_{it}$$
$$(12.2)$$

A detailed definition of the variables is listed in Table 12.4 as well. $T = 1$, 2, ..., 10 is the time trend variable for 1995–2004. To be sure, dummy variables should be added to deal with potential spatial heterogeneity. One way is to divide

---

[5] 1 mu = 1/15 ha. We use "mu" here for the purpose of facilitating the readers to understand the variations of the very small household land holdings in China.

Table 12.4 Summary Statistics and Variables of the Household Data

Definition	Variable	1995				2004			
		Mean	SD	Min	Max	Mean	SD	Min	Max
Total income (yuan)	R	4,291.1	3,202.8	98.3	29,468.4	7,569.3	5,481.3	6.1	102,441.4
Farmland (mu)	X_1	9.5	10.8	0.0	86.0	5.5	6.0	0.0	86.0
Foreland (mu)	X_2	10.5	23.4	0.0	680.0	15.2	24.6	0.0	680.0
Other land (mu)	X_3	0.4	3.2	0.0	95.4	2.5	10.3	0.0	130.0
Labor for land-based activities (day)	X_4	229.1	163.2	0.0	1,120.0	216.4	162.8	0.0	975.0
Labor for other local production (day)	X_5	21.5	65.1	0.0	600.0	27.6	75.7	0.0	600.0
Off-farm employment (day)	X_6	78.2	139.8	0.0	1,200.0	165.6	226.4	0.0	1,500.0
Production expenditure for land-based activities (yuan)	X_7	485.3	403.3	0.0	4,888.3	595.3	552.6	0.0	6,235.6
Area enrolled in the NFPP (mu)	Y_1	0.0	0.0	0.0	0.0	3.9	8.9	0.0	145.5
Area enrolled in the SLCP (mu)	Y_2	0.0	0.0	0.0	0.0	3.4	8.3	0.0	102.3
In Yellow River basin (mu)	Y_{21}	0.0	0.0	0.0	0.0	7.2	14.0	0.0	102.3
In Yangtze River basin (mu)	Y_{22}	0.0	0.0	0.0	0.0	2.1	4.1	0.0	36.5
Area enrolled in the DCBT (mu)	Y_3	0.0	0.0	0.0	0.0	0.9	5.4	0.0	151.6
Area enrolled in the SBDP(mu)	Y_4	0.0	0.0	0.0	1.0	0.00	0.1	0.0	1.0
Inverse of distance to the nearby WCNR area (1/km)	Y_5	0.0	0.0	0.0	0.0	0.1	1.0	0.0	25.0
Education of household head (year)	Z_1	6.1	2.8	0.0	14.0	6.1	2.8	0.0	14.0
Household size (number of people)	Z_2	3.5	1.2	1.0	8.0	3.9	1.4	1.0	9.0
Distance from village to the nearest township (km)	Z_3	7.5	5.5	0.0	25.0	7.5	5.6	0.0	25.0
Average size of forestland per plot (total forestland area/number of forestland plots, mu)	Z_4	1.6	2.0	0.0	24.0	1.4	3.1	0.0	53.0

Table 12.4 (continued)

Definition	Variable	1995				2004			
		Mean	SD	Min	Max	Mean	SD	Min	Max
Average size of farmland per plot (total farmland area/number of farmland plots, mu)	Z_5	3.7	6.9	0.0	75.0	4.4	6.6	0.0	100.0
Topographic dummy (hill = 1; otherwise 0)	Z_6	0.2	0.4	0.0	1.0	0.2	0.4	0.0	1.0
Topographic dummy (mountain = 1; otherwise 0)	Z_7	0.7	0.5	0.0	1.0	0.7	0.4	0.0	1.0
Road condition (hard surface = 1; otherwise 0)	Z_8	0.1	0.3	0.0	1.0	0.1	0.3	0.0	1.0
Communist party membership (yes = 1; otherwise 0)	Z_9	0.1	0.3	0.0	1.0	0.1	0.3	0.0	1.0
Village leader (yes = 1; otherwise 0)	Z_{10}	0.1	0.3	0.0	1.0	0.1	0.3	0.0	1.0

Notes:
1. The numbers were rounded.
2. 1 ha = 15 mu.
3. Labor inputs are the numbers of person-days devoted to production activities.
4. Observations in 1995 and 2004 were 2,137 and 2,232, respectively. There were 556 observations in 1995 and 578 in 2004 in the Yellow River basin; there were 1,581 observations in 1995 and 1,654 in 2004 in the Yangtze River basin.

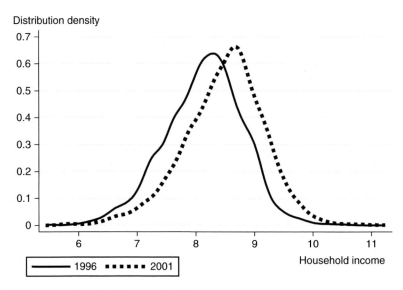

Fig. 12.1 Sample Household Income Distribution

the ten counties in terms of their location in the Yangtze River basin or the Yellow River basin, because the two basins have different natural and socioeconomic conditions. Another way is to include dummies for individual counties. Thus, our model features a fixed-effect specification of panel data, which suggests that household income is a function of production inputs, areas enrolled in the PFPs, and other control variables, including a time trend and regional dummies as well as household and village characteristics.

12.3 Empirical Results

Our fixed-effect empirical model was estimated with the ordinary least square technique. Regression results are summarized in Table 12.5. Four alternative estimations were executed according to our dummy variable selection and the SLCP compensation variation. In the first two regressions, spatial heterogeneity was considered based on whether a household is located in the Yangtze River or Yellow River basin. In the last two regressions, spatial heterogeneity was captured with dummies for individual counties. Under each alternative of regional dummy choice, two scenarios were further considered for the SLCP compensation variation, given that the program adopts different standards of subsidy for the Yangtze River and the Yellow River basin.

Results of the four alternative models are remarkably consistent. While the R^2 values are somewhat low – below 0.4, they are acceptable given the great variations of geography and economy. Among all the production inputs, the coefficients for farmland, other land, labor for other local production, labor for off-farm

Table 12.5 Estimated Results of Income Determination

Variable and coefficient		Alternative model specifications			
		(1)	(2)	(3)	(4)
Farmland	α_1	0.088*** (0.012)	0.0825*** (0.012)	0.075*** (0.012)	0.069*** (0.012)
Forestland	α_2	0.021** (0.010)	0.018* (0.010)	0.008 (0.011)	0.005 (0.011)
Other land	α_3	0.028*** (0.006)	0.025*** (0.006)	0.032*** (0.006)	0.029*** (0.006)
Labor for farming	α_4	0.005 (0.005)	0.004 (0.005)	0.011** (0.005)	0.011** (0.005)
Labor for other local production	α_5	0.055*** (0.003)	0.055*** (0.003)	0.051*** (0.003)	0.052*** (0.003)
Labor for off-farm employment	α_6	0.075*** (0.002)	0.075*** (0.002)	0.075*** (0.002)	0.075*** (0.002)
Production expenditure	α_7	0.052*** (0.004)	0.053*** (0.004)	0.049*** (0.004)	0.049*** (0.004)
NFPP	β_1	0.006* (0.004)	0.007* (0.004)	0.010*** (0.004)	0.011*** (0.004)
SLCP	β_2	0.124*** (0.006)		0.131*** (0.006)	
SLCP – Yellow River basin	β_{21}		0.159*** (0.008)		0.167*** (0.008)
SLCP – Yangtze River basin	β_{22}		0.104*** (0.007)		0.112*** (0.007)
DCBT	β_3	0.029*** (0.009)	0.037*** (0.009)	0.032*** (0.009)	0.039*** (0.009)
WCNR	β_4	1.067*** (0.346)	1.114*** (0.346)	1.538 (0.345)	1.563*** (0.345)
SBDP	β_5	0.025 (0.027)	0.031 (0.027)	0.022 (0.027)	0.029 (0.027)
Years of household head education	γ_1	0.093*** (0.018)	0.093*** (0.018)	0.079*** (0.018)	0.079*** (0.018)
Household size	γ_2	0.245*** (0.026)	0.252*** (0.026)	0.246*** (0.026)	0.256*** (0.026)
Distance to the nearest township	γ_3	-0.028* (0.016)	-0.023 (0.016)	-0.001 (0.016)	0.003 (0.016)
Average size of farmland plots	γ_4	-0.117*** (0.017)	-0.124*** (0.017)	-0.071*** (0.018)	-0.078*** (0.018)
Average size of forestland plots	γ_5	-0.042*** (0.013)	-0.062*** (0.013)	-0.034*** (0.013)	-0.054*** (0.013)

Table 12.5 (continued)

Variable and coefficient		Alternative model specifications			
		(1)	(2)	(3)	(4)
Topography: Hill	γ_6	−0.140*** (0.050)	−0.160*** (0.050)	−0.140** (0.056)	−0.143** (0.056)
Mountain	γ_7	−0.260*** (0.045)	−0.268*** (0.045)	−0.163*** (0.052)	−0.166*** (0.052)
Road	γ_8	0.070** (0.034)	0.077** (0.034)	0.076** (0.033)	0.083** (0.033)
Party member status	γ_9	0.076** (0.035)	0.078** (0.035)	0.098*** (0.035)	0.101*** (0.035)
Village leader status	γ_{10}	0.160*** (0.040)	0.162*** (0.040)	0.132*** (0.040)	0.133*** (0.040)
Time trend	φ	0.047*** (0.001)	0.048*** (0.001)	0.047*** (0.001)	0.047*** (0.001)
Spatial heterogeneity Yangtze vs. Yellow River basin	λ	0.240*** (0.028)	0.280*** (0.029)		
County dummies	λ_{1-9}			Omitted	Omitted
Intercept	α_0	6.938*** (0.074)	6.932*** (0.074)	6.659*** (0.090)	6.631*** (0.090)
R^2		0.341	0.338	0.373	0.369

Note: There were 22,058 observations. *** means significant at the 0.01, level, ** at 0.05 level, and * at 0.10 level; standard errors are in parentheses.

employment, and cash expenditure are significant under all scenarios. The elasticities of these variables range, respectively, from 0.088 to 0.069 for farmland, 0.025 to 0.032 for other land, 0.051 to 0.055 for labor for other local production, 0.075 for labor for off-farm employment, and 0.049 to 0.053 for cash production expenditure. In comparison, forestland and labor for farming have very small positive coefficients that are significant in two of the four regressions each. The elasticity of forestland ranges from 0.005 and 0.021, whereas that of labor for farming ranges from 0.005 to 0.011.

Of the five PFPs included in our regressions, the WCNC has a negative effect on household income at the 99% confidence level. As expected, the closer a household is to a natural reserve, the lower its total income. In contrast, the contributions of the SLCP and the DCBT to total household incomes are the largest – their coefficients range from 0.124 to 0.131 for the SLCP and from 0.029 to 0.039 for the DCBT. The breakdown of the SLCP further shows that it has a larger positive impact in the Yellow River basin, compared to that in the Yangtze River basin, conforming our observation that households in the Yellow River basin had a much larger amount of farmland converted, but their average income level is markedly lower (SFA, 2005). And this observation is reinforced by our finding of the spatial heterogeneity – the total income of a household located in the Yangtze River basin is 24–28% higher than that in the Yellow River basin. Similarly, the coefficients of dummy variables for individual counties are all significant at 95% confidence level or higher. To save space, however, they are not reported here. The coefficient of the NFPP is positive and significant at the 90% confidence level at the least. While it contradicts with the negative impact reported in the literature (Xu et al., 2004; Liu & Zhang, 2006; Du, 2004), it is very small. Finally, the coefficient of the SBDP is insignificant under any of the scenarios.

Our estimated coefficients of the household and village characteristics reveal that years of household head schooling and family size have significantly positive effects on income. The coefficient of the former is 0.079–0.093, and that of the latter is 0.245–0.256. This indicates that the higher the education attainment of the household head and the larger the family size, the greater the household income. On the other hand, the average sizes of the farmland and forestland plots are negatively correlated with household income, with a coefficient ranging from –0.071 to –0.124 for the average farmland plot size and from –0.034 to –0.062 for the average forestland plot size. These results suggest that small and fragmented land holdings are detrimental to income growth. Compared to a village located in a plain, one located in a mountainous or hilly area is disadvantaged in family income generation by a margin of 16–27% and 14–16%, respectively. Moreover, a better road condition makes a 7–8% contribution to household income.

Notably, a household head being a village leader or Communist party member is also beneficial to family income. These indicators of status can result in an income increase of 13–16% and 7–10%, respectively. Also, the strongly positive coefficient of time trend indicates that technical and institutional changes have facilitated income growth by almost 5% a year. The coefficient of distance from a village to its nearest township has the correct negative sign, but it is insignificant.

12.4 Conclusions and Discussion

This study was motivated by our observation that while it is of broad interest and great relevance to compare and contrast the income impacts of China's Primary Forestry Programs, little has been done along this direction of research. So, the first thing we did was to build a large panel dataset based on an appropriate sampling and enumerating approach. To that end, we gathered survey information from more than 2,100 households in ten counties and for a period of 10 years (1995–2004). In addition to household income, included in our data are an exhaustive number of variables of production inputs, program engagements, and household and village features. Then, we estimated a fixed-effect model of multiple specifications to investigate the income determinants, including the PFPs. These steps have allowed us to derive a rich and interesting set of empirical results.

First, it is found that the ecological restoration and resource development programs have affected household incomes in different ways. While the WCNR has had a negative impact on household income, three other programs – the SLCP, the DCBT, and the NFPP – have made positive contributions. Among them, the SLCP has the largest effect on household income, followed by the DCBT. These highly significant effects are consistent with both the heavy government subsidies and the findings by others (Uchida et al., 2005; Yao, Guo, & Huo, 2008). While these effects were anticipated, the positive effect of the NFPP contradicts what has been reported in the literature (Xu et al., 2004; Wu & Liu, 2002). It is true that at the beginning of the NFPP, sudden logging bans and forestry activity contractions inflicted a lot of hardship on the local farmers. However, later the government hired local farmers for forest protection and management and relaxed timber harvest restrictions in non-state forests that are subject to the NFPP coverage.

More importantly, the NFPP has triggered an adjustment to the local economic structure, including moving away from forestry-related production and from local employment. Now, income generated from off-farm activities accounted for about a half in the sample counties of Sichuan and 40% in the counties of Shaanxi. And the time period of our data is longer than that used in other studies, permitting us to capture the induced income effect in more recent years. So, the positive but small effect of the NFPP seems plausible.

Additionally, the SBDP has little impact on household income, which is reasonable given the smaller amount of public investment that is scattered over a broad area. From the perspective of poverty reduction and livelihood enhancement, this finding suggests that the government may rethink its practices – whether and how the program can be more concentrated and better executed. As to the WCNR, a key question is how the government can more effectively integrate nature conservation with economic development to mitigate the negative impact of excluding or restraining local farmers' access to nature reserves. Certainly, it is our view that the effectiveness of the SLCP, the DCBT, and the NFPP can and should be improved as well.

Overall, the contributions of various production inputs are in line with our expectations. Notably, farmland has a large impact on income because farming remains the major source of income. Meanwhile, it appears that other land for home gardening, while small in amount, has a large impact on income, given its intensified use in growing more profitable cash crops. Likewise, labor for off-farm employment plays a major role in family income generation. These results imply that more attention should be directed to increasing income from cash crops and off-farm employment. The significant effect of the time trend variable shows that technical and institutional changes have contributed to household income growth. In conjunction with our finding of household head schooling, this suggests that education, professional training, and extension service should be strengthened in the future. On the other hand, because of the limited farmland, the abundantly available work time for farming does not contribute to income generation at the margin, confirming what has been extensively documented (e.g., Liu & Yin, 2004). Likewise, forestland makes a very small contribution to income. While the share of forestry income (including subsidies of PFPs) has also increased from 0.9% in 1999 to 1.9% in 2006 nationwide, it remains a fairly tiny component of household income (CNSB, 2007). Clearly, how to increase farmland and forestland productivity is a challenge to continued income growth and livelihood improvement.

Also as expected, household size has a positive impact on income, which is a reflection of the fact that the original land devolution was based partially on household size. The larger the household, the more land it got. Of course, the larger the household, the more laborers it has as well. But because of land fragmentation caused by how the household responsibility system was implemented a quarter century ago, it is unsurprising to find that the average sizes of farmland and forestland per plot are negatively correlated with household income. Similar evidence is also reported by Chen and Brown (2001). This indicates a strong need to address the question of how to consolidate land uses and improve their efficiency and productivity. While the Chinese government has been making efforts along this direction as part of its strategy of rural development, our results reinforce their importance.

References

Chen, K., & Brown, C. (2001). Addressing shortcomings in the household responsibility system: an empirical analysis of the dual farmland system in Shandong Province, *China Economic Review, 12*, 280–292.

Chen, S. H., & Ravillion, M. (2004). *How have the World's poorest fared since the early 1980s.* World Bank Discussion Paper, WPS 3341.

China National Bureau of Statistics (2002–2007) *China statistical yearbook 2002~2007.* Beijing: China Statistics Press (in Chinese).

Du, S. H. (2004). *Participatory management and the protection of farmer's rights and interests.* Beijing: China Economic Press.

FAO. (2006). *Global forest resources assessment 2005.* Rome: The Food and Agricultural Organization.

Hu, C. (2002). *A survey report of closing hillsides to facilitate afforestation and interests.* Beijing: China Economic Press (in Chinese).

Li, Z. (2001). Evaluation of policies for harnessing the land of China. *Chinese Rural Economy, 9*, 35–39.

Liu, C. et al. (2005). Regional socioeconomic and ecological effects of NFPP in China. *Journal of Ecology, 3*, 428–434.

Liu, C., & Lü, J. Z. (2008). *National prospective of the effect priority forestry programs on farmers' income during 1998–2006*. Working Paper, China National Forestry Economics and Development Research Center.

Liu, C., & Yin, R. S. (2004). Poverty dynamics as revealed in the production performance of rural households: the case of west Anhui, China. *Forest Policy and Economics, 6*, 391–401.

Liu, C., & Zhang, W. (2006). The impact of environmental policy on household income and activity choice: Evidence from sand control program for areas in the vicinity of Beijing and Tianjin. *Journal of Economics (Quarterly) China, 1*, 273–290.

Ni, J., Wang, Y. P., & Yang, Z. W. (2002). Evaluation of the natural forestry protection project in Yunnan. *Forestry Inventory and Planning, 1*, 34–38 (in Chinese).

State Forestry Administration. (2002–2005). *China forestry development report*. Beijing: China Forestry Publishing House (in Chinese).

Schultz, T. W. (1964) *Transforming traditional agriculture*. Yale: Yale University Press.

Uchida, E., Xu, J. T., & Rozelle, S. (2005). Grain for green: Cost-effectiveness and sustainability of China's conservation set-aside program. *Land Economics, 81*(2), 247–264.

Uchida, E., Xu, J. T., Xu, Z. G., & Rozelle, S. (2007). Are the poor benefiting from China's land conservation program? *Environment and Development Economics, 12*(4), 593–620.

UNDP (2005). *Human development report 2005*. New York: United Nations.

Weyerhaeuser, H., Wilkes, A., & Kahrl, F. (2005). Local impacts and responses to regional forest conservation and rehabilitation programs in China's northwest Yunnan Province. *Agricultural Systems, 85*, 234–253.

World Bank. (2000). *Towards a revised forest strategy for the World Bank Group*. Washington, DC: The World Bank.

Wu, S., & Liu, C. (2002). Environmental and economic evaluation of natural forestry protection program. *Forestry Economics, 11–12* (in Chinese).

Xu, J. T., Tao, R., & Xu, Z. G. (2004). Conversion cropland to forest and grassland program: Cost-effectiveness, structural effect and economic sustainability. *Chinese Academy of Science, 1*, 139–161.

Xu, Q. (2003). The development system of community forestry and the study of converting cropland to forests – take Shanxi Province and Shun Country as an example. *Forestry Economics,* (supplement) (in Chinese).

Yao, S. B., Guo, Y. J., & Huo, X. X. (2008). *An empirical analysis of effects of China's land conversion program on farmers' income growth and labor transfer*. Northwest A&F University College of Economics and Management working paper (in Chinese).

Yin, R. S., Xu, J. T., Li, Z., & Liu, C. (2005) China's ecological rehabilitation: The unprecedented efforts and dramatic impacts of reforestation and Slope Protection in Western China. *China Environment Series, 6*, 17–32.

Yin, R. S., & Yin, G. P. (2008) *China's ecological restoration programs: Initiation, implementation, and challenges*. Michigan State University Department of Forestry working paper.

Zhi, L., & Shao, A. Y. (2001) Practice and considerations of converting cropland to forest and grassland. *Forestry Economics, 3*, 43–46 (in Chinese).

Chapter 13
Agricultural Productivity Changes Induced by the Sloping Land Conversion Program: An Analysis of Wuqi County in the Loess Plateau Region

Shunbo Yao, Hua Li, and Guangquan Liu

Abstract The goal of this chapter is to exam the agricultural productivity change induced by the Sloping Land Conversion Program (SLCP), using the Malmquist index method and household data collected from Wuqi. We find that during the period of 1998–2004, the total factor productivity (TFP) grew by 15.8%. While numerous households suffered a TFP decline, the majority of them experienced a large gain. By decomposing the TFP, we further show that its increase is due exclusively to the improvement of technical efficiency, rather than to technological change. To validate these findings and put them in perspective, we also estimated the TFP change with county-level aggregate data. It is revealed that driven by technological change and scale efficiency, the TFP grew slightly during the period of 1992–1998; because of the tremendous cropland reduction and production mode shift caused by implementing the SLCP, the TFP declined substantially during the first 3 years of the program; due to continued improvement of technical efficiency, however, its growth accelerated later. Altogether, our evidence consistently shows that implementing the SLCP has contributed to the agricultural TFP growth in the longer term, and that the efficiency improvement has resulted mainly from the increased public expenditures for extension services and diffusion of technical knowledge. Wuqi's experience proves that it is possible to achieve environmental conservation and productivity increase simultaneously, even facing a huge cropland reduction and production mode alternation.

Keywords Land conversion · Total factor productivity · Malmquist index method · Technical efficiency · Technological change · Scale economy

---

S.Yao (✉)
College of Economics and Management, Northwest A&F University, Yangling, Shaanxi 712100, P.R. China
e-mail: yaoshunbo@126.com

13.1 Introduction

In 1999, the Chinese government initiated the Sloping Land Conversion, or Grain for Green, Program (hereafter, SLCP) to retire and rehabilitate farmland that has been degraded primarily as a result of soil erosion, induced by water runoffs and wind damages (Yin, Xu, Li, & Liu, 2005). As the largest ecological restoration program in the developing world, the SLCP aims to reverse the environmental deterioration while transforming the structure of the rural economy and improving farmers' livelihoods in western China (SFA, 2003). Even though there have been studies of its socioeconomic impacts, few have analyzed the changes in agricultural productivity induced by implementing the SLCP. The objective of this chapter is to fill this void in the literature.

Several articles have provided a clear and complete description of the SLCP in terms of its geographic coverage, technical measures, participation subsidies, and implementation effectiveness (e.g., Yin & Yin, 2008; Bennett, 2007; Xu, Yin, Li, & Liu, 2006). Moreover, many scholars have investigated its impacts on income, employment, cost efficacy, and food supply. Using household survey data, Uchida, Xu, Xu, and Rozelle (2005) found that 40% of the plots enrolled in the program had a yield lower than the level of compensation, implying that there was a significant degree of over compensation. Based on the purchase power parity, they further demonstrated that the average compensation is 50% higher than the budgetary outlay of the Conservation Reserve Program (CRP) in the US. These results indicate that the Chinese government might be able to generate fiscal savings if the payments could be made to more accurately reflect the variation in the opportunity costs of different plots.

Using similar difference-in-differences models and household surveys, Xu, Tao, and Xu, (2004) showed that over the period of 1999–2003, the growth rates in average income varied across regions, but the overall impact of the SLCP on participants' income is statistically insignificant. In contrast, based on panel data and a fixed-effect model, Liu and Zhang (2006) detected a positive impact of converting farmland to forestland on household income in the vicinity of Beijing and Tianjin. According to their estimation, household income would be 17.4% higher if the sample village had introduced the program 1 year earlier. Using quartile regressions, Zhang, Swanson, and Kontoleon (2005) confirmed that the SLCP is making a significantly positive impact on the incomes of poor farmers. In another study, Xu, Yin, and Zhou (2007) illustrated that the SLCP has contributed to the social transformation of the traditional rural society by enabling the workers freed up from farming to seek off-farm jobs in or outside of the locale.

Additionally, scholars have examined the potential effect of the program on food security. Using a multi-objective programming model, Feng, Yang, Zhang, Zhang, and Li (2005) simulated the impact of the SLCP on China's grain supply in the upper reaches of the Yangtze River and the Yellow River. They found that this impact was in the range of 2–3%, suggesting that the SLCP might not have a major effect on China's grain supply. But the impact in certain local areas can be significant. Under modified assumptions regarding farmers' production behavior, such as their response to price changes, Xu et al. (2006) revealed that the SLCP has an even smaller effect on China's grain production and little influence on prices or food imports. These results thus suggest that implementing the SLCP has not led to a decline of the agricultural production that is in proportion to the farmland reduction. Nevertheless, it remains unclear whether and to what extent implementing the SLCP has potentially affected agricultural production. Therefore, this study is motivated to address these questions empirically.

Of course, as a result of retiring degraded croplands and converting them into forest and grass coverage, the land base for farming is reduced and the production mode is altered. However, the reduction of farmland and the alteration of production mode should not inevitably lead to a proportionate decline of the agricultural production if a combination of different components of productivity – technical efficiency, technological change, and scale economy (Färe & Grosskopf, 1992) – can offset the negative effects caused by implementing the SLCP. If the agricultural productivity has indeed been adversely impacted, then it will have undesirable consequences to food security, farmer income, and rural development (Lin, 1992; Liu, 2003), which will in turn call into question the sustainability of the SLCP and other related efforts of environmental conservation. Otherwise, if implementing the program has significantly improved the agricultural productivity, or at least it has not resulted in a decline of the agricultural productivity, then its design and execution are validated – environmental improvement and economic growth can be accomplished simultaneously. Further, it is known that even if there has been a significant increase in the aggregate agricultural productivity along with implementing the SLCP, the different productivity components may have witnessed a variation in their individual paths of change. It is also anticipated that not all of the sample units (rural households in this case) have experienced a uniform pattern of performance shift.

In this chapter, we use the Malmquist index method and Wuqi county in Shaanxi as a study site to estimate the changes in agricultural productivity and its components induced by implementing the SLCP. Our primary data feature a large sample of rural households in 1998, before the SLCP was initiated, and in 2004, when the local land conversion was virtually completed. It is found that in contrast to its earlier slow rate of growth, the total factor productivity (TFP) accelerated during the period of 1998–2004. Also, although numerous households suffered a TFP decline, a majority of them experienced a TFP gain. By decomposing, we further show that the TFP increase is due exclusively to the improvement of technical efficiency, rather than to technological change. Compared to their levels in 1998, technical efficiency improved by 51.9%, whereas technological change declined by 23.8%.

The existence of technological regress indicates that implementing the SLCP indeed caused a shock to the existing technology and a transformation of the production mode. However, farmers could increase their productivity by improving their efficiency of input use, with extension services and information diffusion provided by local public agencies. Thus, it is possible to achieve environmental improvement and productivity increase at the same time, even facing cropland reduction and production mode alternation. The chapter is organized as follows. Before presenting our empirical results, we will discuss our method and data used for measuring the TFP

change. Finally, we will make some closing remarks and discuss the implications of
our work.

13.2 Method

Productivity is traditionally measured in terms of a single input factor, such as labor
productivity and land productivity. In reality, however, all productive factors – labor,
land, capital, and others – are utilized simultaneously, and substitution exists among
them. As such, TFP – a productivity measure involving all factors of production – is
preferred over a partial-factor productivity measure (Coelli, Rao, & Battese, 1998).
Simply put, the TFP of a firm is the ratio of the outputs that it produces to the inputs
that it uses. As already noted, TFP includes three components: technical efficiency,
scale economy, and technological change.

There have been numerous TFP studies. Some analyzed the efficiency and productivity change of large agricultural enterprises during their transition to a market
economy (e.g., Lissitsa & Odening, 2005), while others investigated the sources of
sectoral growth in an economy (e.g., Gopinath & Roe, 1997). In the Chinese context, most of the studies examined productivity growth over different periods of its
modern history (e.g., Yi, Fan, & Li, 2003), over different regions (e.g., Xu, 2005),
or the composition of different productivity components (e.g., Zhang & Shi, 2003;
Huang & Rozelle, 1996). Despite its policy relevance and urgent need, however,
little has been done to explore the gross and compositional productivity shifts in
agriculture, induced by implementing the SLCP. This is surprising in view of the
broad geographic scope of the program and its extensive conversion of croplands.

The Malmquist index (MI) method is a nonparametric measurement of productivity, and it is defined using distance functions, which allow one to describe a
multiple-input, multiple-output production technology without the need to specify
a behavioral objective (i.e., cost minimization or profit maximization). Further, distance functions are distinguished by orientation. An input distance function defines
the maximal proportional contraction of the input vector $(x_1 \ldots x_N)$, given an output
vector $(y_1 \ldots y_M)$. An output distance function considers a maximal expansion of
the output vector, given the input vector.

Assume that there are K firms (households in this case), the output distance function at period t is defined as:

$$D_o(x_t, y_t) = \inf \{\theta \ (x_t, y_t/\theta) \in I_t, \theta \geq 0\} \tag{13.1}$$

where θ is a scalar, $x_t \epsilon\ R^{N+}$ and $y_t \epsilon\ R^{M+}$ are the input and output vectors at period t,
and I_t is the production possibility set, representing the set of all outputs, y_t, that can
be produced with input vector, x_t. The subset of I_t that features the maximum outputs
produced from a given combination of inputs is called the production frontier. As
shown in Fig. 13.1, the movement of a firm's production in relation to the frontier
over time is called efficiency change, the movement of the frontier itself is called

Fig. 13.1 Components of Productivity Change with Variable Returns to Scale

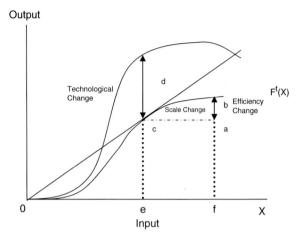

technological change, and the change in scale of operation is called scale economy (Coelli et al., 1998; Managi et al., 2004).

The MI measures the TFP change between two data points by calculating the ratio of the distances of each data point relative to a common technology. Because the reference technology can be that of period s (the base period) or t (the current period), there are two alternative productivity indices – one based on the base period technology (M_s) and the other on the current period technology (M_t). Often, empirical analyses take the geometric mean of the two to obtain a more stable result $MI(y_s, x_s, y_t, x_t) = [M_s \times M_t]^{1/2}$ (Färe & Primont, 1995). The output-oriented Malmquist TFP index between period s and period t is then given by

$$MI(y_s, x_s, y_t, x_t) = [M_s \times M_t]^{1/2} = \left[\frac{d_o^s(x_t, y_t)}{d_o^s(x_s, y_s)} \times \frac{d_o^t(x_t, y_t)}{d_o^t(x_s, y_s)} \right]^{1/2} \qquad (13.2)$$

where $d_o^s(y_t, x_t)$ is the distance from period t observation to period s technology. A value of MI greater than one will indicate positive TFP growth from period s to period t, whereas a value less than one indicates a TFP decline. Note that the first TFP index in the brackets is evaluated with respect to period s technology and the second with respect to period t technology.

No technical inefficiency is incorporated into Equation (13.2). In case there exists technical inefficiency, the above equation becomes:

$$MI(y_s, x_s, y_t, x_t) = \frac{d_o^t(y_t, x_t)}{d_o^t(y_s, x_s)} \times \left[\frac{d_o^s(x_t, y_t)}{d_o^s(x_s, y_s)} \times \frac{d_o^t(x_t, y_t)}{d_o^t(x_s, y_s)} \right]^{1/2} \qquad (13.3)$$

where the ratio outside the brackets measures the change in the output-oriented technical efficiency (TE) between period s and period t. If the TE change is greater than one, then it suggests an efficiency improvement; otherwise, it indicates an efficiency deterioration. Also, the TE term can be further decomposed into "pure" efficiency and scale efficiency (Coelli et al., 1998). Given the potential variability of operation scales, this decomposition is warranted if possible. The whole term inside the brackets reflects technological change (TC). A TC value greater than one means a technological progress; otherwise, it indicates a technological regress.

To estimate the MI, our first task is to obtain the four distance measures by solving the following linear programming problems:

$$\left[d_o^1\ (y_t, x_t) \right]^{-1} = \max_{\phi, \lambda}\ \phi,\ s.t.\ -\varphi y_{it}\lambda + Y_t\lambda \geq 0,\ X_{it}\lambda - X_t\lambda \geq 0,\ \lambda \geq 0, \tag{13.4}$$

$$\left[d_o^s\ (y_s, x_s) \right]^{-1} = \max_{\phi, \lambda}\ \phi,\ s.t.\ -\varphi y_{is}\lambda + Y_s\lambda \geq 0,\ X_{is}\lambda - X_s\lambda \geq 0,\ \lambda \geq 0, \tag{13.5}$$

$$\left[d_o^t\ (y_s, x_s) \right]^{-1} = \max_{\phi, \lambda}\ \phi,\ s.t.\ -\varphi y_{is}\lambda + Y_t\lambda \geq 0,\ X_{is}\lambda - X_t\lambda \geq 0,\ \lambda \geq 0, \tag{13.6}$$

and

$$\left[d_o^s\ (y_t, x_t) \right]^{-1} = \max_{\phi, \lambda}\ \phi,\ s.t.\ -\varphi y_{it}\lambda + Y_s\lambda \geq 0,\ X_{it}\lambda - X_s\lambda \geq 0, \lambda \geq 0, \tag{13.7}$$

It should be noted that in order to determine whether implementing the SLCP has contributed to agricultural productivity growth, a counterfactual must be established to detect what would have happened if the program had not been initiated. Because most of the households in Wuqi were enrolled in the program, we do not have a sufficient number of non-participating households to build this baseline. Also, because the earliest year of our household survey is 1998 – right before the program was launched, we do not have adequate sample information to identify what might be the multi-year TFP growth trend prior to that time as well. The only available option is thus to estimate the county's TFP over an extended period based on the aggregate data compiled by its Statistics Bureau. While this may not be ideal, it should still provide a plausible baseline nonetheless.

13.2.1 Data

As indicated, our primary dataset for evaluating the agricultural productivity changes induced by implementing the SLCP came from household surveys in Wuqi county of northern Shaanxi (Fig. 13.2). Wuqi is typical of the Loess Plateau region, where the landform is full of hills and gullies of yellow soil. In this arid region, agriculture is largely rain-fed. Due to the lack of precipitation, however, crop yields were abysmally low historically. Prior to 1998, the cultivated area in Wuqi reached 123,700 ha, of which only four percent had a yield above 2,250 kg/ha. To increase income, 17% of the households also raised goats, whose population was as high as 280,000. As a consequence of extensive farming and open grazing, the ground

Fig. 13.2 Location of the Study Site

vegetation was heavily degraded, worsening the erosion of the soft silty soil found there. By the end of 1998, 3,696 km² of the county's 3,791.5 km² suffered erosion, with 153,000 tons of topsoil washed downstream per square kilometer (Wuqi Statistics Bureau, 2003).

In responding to this environmental disaster, Wuqi pioneered retiring croplands on steep slopes and converting them to forest and grass coverage on a massive scale, beginning in 1999. Croplands were reduced to 10,000 ha (Wuqi Statistics Bureau, 2003), and farmers were encouraged to adopt new farming practices and improved seeds. With financial and technical assistance provided by the local public agencies, in addition to the subsidies made by the central government, land use became more intensive. Meanwhile, the traditional open herding practice was banned; goats, often introduced, improved breeds, were required to be kept in pens. As a result, agricultural production did not suffer a huge decline (Table 13.1); indeed, crop yields per ha increased markedly. Even though the cropland holding per household dropped

Table 13.1 Arable Land and Crop Yield in Wuqi Farm Households

Year	Average arable land (ha/family)	Corn		Potato		Minor crops	
		Average sown area (ha/family)	Yield (kg/ha)	Average sown area (ha/family)	Yield (kg/ha)	Average sown area (ha/family)	Yield (kg/ha)
1998	3.40	0.17	3,350.55	0.83	8,076.15	2.40	1,160.4
2004	0.76	0.09	4,421.55	0.33	9,974.25	0.34	1,390.2

Notes: The data were from authors' surveys in August 2005; Minor crops refer to millet, buckwheat, and beans.

from 3.40 ha in 1998 to 0.76 ha in 2004, for instance, the yields of corn and potato increased from 3,356 to 4,422 kg/ha and 8,076 to 9,974 kg/ha, respectively.

Furthermore, the retired cropland soon became green, which can even be detected from satellite images. The county's vegetation cover grew from 19.2% in 1998 to 62.9% in 2006 (Wuqi SLCP Office, 2007). For this reason, Wuqi's land retirement and conservation initiatives have attracted national attention, and thus Wuqi has been designated as an SLCP demonstration site and soil and water conservation model by the national agencies.

Therefore, scrutinizing Wuqi's implementation of the SLCP and the induced changes in its economy and environment is of great relevance to better understanding the effectiveness of the program in the entire Loess Plateau region. For that purpose, the period between 1998 and 2004 was representative, because it is one in which the local households witnessed a major shift in their agricultural production, with almost all of them quickly enrolled in the ecological restoration program and its targets virtually completed by 2004.

We conducted our household surveys in 2005, with a total of 1,621 households in 12 townships being randomly interviewed. Having removed those households with incomplete or inaccurate information, 1,340 were left for this study. The output and input variables of the sample households are given in Tables 13.2 and 13.3. Output variables include the output value of crop production and the output value of livestock product, which were aggregated using their 1990 constant prices. Crop production refers to the production of corn, potatoes, and other minor crops (e.g., millet, buckwheat, and beans), and livestock production refers to the production of goats and pigs. Input variables include cultivated land, labor for farming and animal husbandry, fertilizers (the sum of N, P, K, and composite fertilizers), manure, crop coversheets, seeds, feedstock, and fodder.

While the land, labor, and fertilizer variables are defined straightforwardly, the manure, crop coversheet, feedstock, and fodder variables warrant a brief explanation. One study suggests that the nutrient content of manure is roughly 2.2%, and its rate of use is 75% (Chen, 2006). The amount of manure is thus defined as a product of the physical weight and 0.0165. Crop coversheets are the total weight of the plastic sheets (kg) used for enhancing topsoil moisture and temperature once

Table 13.2 Agricultural Outputs of Sample Households in 1998 and 2004. Unit: kg/household, head/household

Year	Variable	Corn	Potato	Minor crops	Goat	Pig
1998	Mean	928	3,580	2,625	19.67	0.84
	Maximum	10,400	70,000	3,700	150	300
	Minimum	10	0	0	0	0
2004	Mean	705	2,851	537	6.08	7.97
	Maximum	18,750	15,000	4,500	250	4,700
	Minimum	40	75	0	0	0

Note: The data were from authors' surveys in August 2005.

Table 13.3 Agricultural Inputs of Sample Households in 1998 and 2004. Unit: day, piece, hour, kg/household

Year	Variable	Labor use	Fertilizer use	Manure	Crop cover sheets	Improved seeds	Feed stock	Fodder
1998	Mean	722	1,346	9,220	0	0	733	627
	Max	2,640	7,350	45,000	0	0	5,000	25,000
	Min	300	0	0	0	0	0	0
2004	Mean	543	795	3,554	7	1.1	713	4,208
	Max	1,080	5,810	30,500	60	107.0	6,500	50,000
	Min	200	80	10	0	0	0	0

Note: The data were from authors' surveys in August 2005.

seeds or seedlings are planted. Feedstock is the total weight of corn and other grains used to feed goats and pigs, and fodder is the total weight of collected grasses and crop straws. To validate our findings and put them in perspective, we also estimated Wuqi's agricultural TFP with the county-level aggregate statistical data for the period of 1991–2004, the output and input variables of which are similar to those defined in our household surveys.

Tables 13.2 and 13.3 also list the summary output and input statistics of the sample households, including their mean and minimum and maximum values. Table 13.2 shows that with a significant reduction of arable land induced by the SLCP, an average household's annual corn and potato outputs were reduced by only 24.03% and 20.36%, respectively, in 2004, compared to those in 1998. In contrast, its total output of other minor crops decreased by 79.54% in 2004, compared to that in 1998. Likewise, goats raised per household in 2004 were 77.99% lower than those in 1998. The number of pigs raised per household was 84.81% higher in 2004 than that in 1998.

Accordingly, with the substantial decrease of arable land resulted from implementing the SLCP, the fertilizer use rate was reduced by 40.94% in 2004 than that in 1998, and the application of manure decreased by 22.84%. Farmers had rarely used cropland coversheets and improved seeds before the program was introduced, but these inputs were extensively used by 2004. In 2004, feedstock per household was reduced by 2.73%, in comparison to that in 1998, while fodder consumption increased by 571.13%.

13.3 Results

As shown in Table 13.4, the agricultural TFP in Wuqi grew 15.8% during the period of 1998–2004, leading to an annual growth rate of almost 2.5%. This demonstrates that implementing the SLCP may have contributed to the overall agricultural productivity growth. More specifically, the technical efficiency improved by 51.9%, whereas the technological change declined by 23.8%. The substantial improvement in technical efficiency is fully attributable to "pure" efficiency enhancement. This

Table 13.4 The Growth of Agricultural TFP and Its Components in Wuqi During 1998–2004, Estimated with Household Survey Data

TFP	TC	TE	PE	SE
1.158	0.7623	1.519	1.556	0.9762

Note: The software used for estimation is TFPIP (Version 1.0, developed by Tim Coelli, Centre for Efficiency and Productivity Analysis, University of New England, Australia).

is so even with a slight decrease in scale efficiency. In 2004, "pure" efficiency was 155.6% of its level in 1998, whereas scale efficiency dropped to 97.6% of its level in 1998. These findings suggest that productive factors were utilized a lot more efficiently, triggered by the retirement and conversion of cropland.

To validate our findings and put them in perspective, we also estimated Wuqi's agricultural TFP and its components, as promised, with the county-level aggregate statistics for the period of 1991–2004. Table 13.5 reports that the TFP grew slowly before the SLCP was initiated, and then it declined during the first 3 years of the program implementation (1999–2001); however, its growth accelerated tremendously in the last 3 years of our data coverage (2002–2004). Obviously, these outcomes corroborate our results derived from the household survey data. They also illustrate that the slow TFP growth prior to launching the SLCP is attributable to both technological change and scale economy. Thereafter, however, the production experienced such a great shock by the SLCP that the basic conditions of agricultural scale and technology must be reconfigured. As a consequence, the TFP declined in the following 3 years. Fortunately, technical efficiency kept improving, which has served to mitigate the TFP decline first and then to accelerate its growth more recently.

Table 13.5 The Growth of Agricultural TFP and Its Components in Wuqi During 1991–2004, Estimated with County-Level Aggregate Statistics

Year	TE	TC	PE	SE	TFP
1991	0.9595	1.0413	0.9748	0.9843	0.9991
1992	0.9593	1.0511	0.9741	0.9848	1.0083
1993	0.9763	1.0513	0.9647	1.0120	1.0264
1994	0.9677	1.0611	0.9443	1.0248	1.0268
1995	0.9634	1.0413	0.9048	1.0648	1.0032
1996	0.9937	1.0355	0.9064	1.0963	1.0290
1997	1.0422	1.0455	0.9963	1.0461	1.0897
1998	0.9975	1.0152	0.9463	1.0541	1.0127
1999	0.9537	0.9453	1.1180	0.8530	0.9015
2000	1.0739	0.8955	1.4280	0.7520	0.9616
2001	1.1237	0.8864	1.5083	0.7450	0.9960
2002	1.2827	0.8965	1.5180	0.8450	1.1499
2003	1.3787	0.8498	1.5386	0.8961	1.1717
2004	1.5110	0.7920	1.5889	0.9510	1.1967

Note: The software used for estimation is TFPIP (Version 1.0, developed by Tim Coelli, Centre for Efficiency and Productivity Analysis, University of New England, Australia).

These remarkably consistent findings have lent us confidence in our estimation and decomposition of Wuqi's agricultural TFP changes over time, and in their relationship with the cropland retirement and conversion program.

Listed in Table 13.6, the distribution of household TFP growth is of interest as well. A great deal of performance variation exists among the households. There were 196 households whose TFP decline was more than 50%, 292 with a TFP decline in the range of 15–50%, and 122 with a TFP decline less than 15%. Meanwhile, 300 households had a TFP growth less than 50%, and 142 had a TFP growth in the range of 50–100%, 152 in the range of 100–200%, and 136 in the range of greater than 200%. Overall, 54.5% of the 1,340 households experienced a TFP growth, which more than offset the TFP decline experienced by other households.

With respect to technological change, 454 households experienced a decline of greater than 50%, 489 declined by 15–50%, and 110 by less than 15%; however, 155 experienced a gain of less than 50%, 51 grew by 50–100%, and 81 grew by greater than 100%. Put differently, more than 78% of the households suffered a decline in technological change. Altogether, 344 households suffered a decline in technical efficiency, of which 104 had a drop of more than 50%, 166 dropped by 15–50%, and 74 declined by less than 15%. Meanwhile, 255 households had a growth rate of less than 50%, 233 had a growth rate in the range of 50–100%, 248 were in the range of 100–200%, and 270 grew more than 200%. As a result, 74.3% of the 1,340 households improved their technical efficiency. Similarly, 234 households had a "pure" efficiency gain of less than 50%, 199 had a gain in the range of 50–100%, and 518 had a gain in the range of greater than 100%. On the other hand, 111 had a decline of at least 50%, 179 had a decline of 15–50%, and 99 had a decline of less than 15%. Twenty-two households had an SE decline of more than 50%, 98 had a decline between 15 and 50%, and 631 had a decline below 15%; 525 households had an SE increase of less than 50%, and 64 had an SE increase of over 50%.

Table 13.6 The Distribution of TFP Growth of Individual Households During 1998–2004

Growth rate	TFP		TC		TE		PE		SE	
	No.	%	No.	%	No.	%	No.	%	No.	%
≤50%	196	14.63	454	33.88	104	7.76	111	8.28	22	1.64
(−50%, −15%]	292	21.79	489	36.49	166	12.39	179	13.36	98	7.31
(−15%, 0]	122	9.10	110	8.21	74	5.52	99	7.39	631	47.09
(0,50%]	300	22.39	155	11.57	255	19.03	234	17.46	525	39.18
(50%, 100%]	142	10.60	51	3.81	223	16.64	199	14.85	24	1.79
(100%, 200%]	152	11.34	46	3.43	248	18.51	242	18.06	19	1.42
≥200%	136	10.15	35	2.61	270	20.15	276	20.60	21	1.57
Sum	1,340	100.0	1,340	100	1,340	100	1,340	100	1,340	100

Note: TFP = total factor productivity, TC = technological change, TE = technical efficiency, PE = "pure" efficiency, and SE = scale efficiency.

13.4 Conclusions and Discussion

Using the Malmquist index method and household survey data collected from Wuqi county of northern Shaanxi, we have assessed the TFP change in agriculture, triggered by implementing the SLCP. It is found that the overall TFP grew by 15.8% during the period of 1998–2004; and our alternative estimation with aggregate statistics indicates even a faster average TFP growth for the same period. Meanwhile, it reveals that the TFP growth was slow prior to the time when the SLCP was launched (1991–1998). While implementing the program depressed the agricultural TFP significantly during the first 3 years, it boosted the TFP growth thereafter at an unprecedented pace. Altogether, our evidence strongly suggests that implementing the SLCP has contributed to the overall agricultural productivity growth in the longer term.

By decomposing the TFP, our analyses further show that the recent TFP increase is due exclusively to the significant improvement of technical efficiency, instead of technological progress. Compared to their levels in 1998, technical efficiency improved by 51.9%, whereas technological change dropped by 23.8%. The coexistence of efficiency enhancement and decline in technological change indicates that implementing the SLCP did cause a change of production mode and thus a technological shock, but a majority of farmers could maintain and even increase their productivity by improving their technical efficiency. Additionally, our work indicates that the main reason why the technical efficiency became higher is that the "pure" efficiency was greatly boosted thanks to much better allocation of production factors, following the initiation of the SLCP. This is so despite the reduced scale efficiency.

The above findings carry significant intellectual merit and policy relevance. First, they demonstrate that it is possible to achieve environmental improvement and productivity increase at the same time, even facing a huge cropland reduction and production mode alternation. To that end, one crucial step is to improve the technical efficiency continuously. Second, while it has rarely been pursued in the realm of China's farmland rehabilitation, this type of micro econometric study of agricultural productivity sheds significant new light in terms of how the production structure and process have been affected and what can be done to mitigate the negative impacts of ecological restoration and to promote productivity growth.

Nevertheless, the question remains why and how the production factor allocation could have become more efficient over recent years. Our field observations led us to think that it has much to do with the successful efforts of extension services and knowledge dissemination in, for example, introducing new varieties of crop and goat, concentrating on the fewer types of crops with higher yields, improving the breeding and feeding practices, and expanding the use of cropland coversheets. And these steps were induced by the SLCP and assisted by the local public agencies. To test these hypotheses, we conducted a simple regression, in which the dependent variable is the estimated TFP scores of individual households whereas the independent variables include the ratio of the area of converted sloping cropland to total area of cropland, and the growth rates of the expenditure for extension services and diffusion of technical knowledge as well as labor and capital use per ha.

Table 13.7 Regression
Results of the TFP Drivers
During 1998–2004

Variable	Coefficient
Constant	2.55**
Growth rate of labor use per ha	0.21**
Growth rate of capital use per ha	0.16**
Ratio of the converted sloping cropland to total cropland area	0.49***
Growth rate of the expenditure for extension services and knowledge diffusion per ha	0.92***
F	31.88
R^2	0.86

Note: The software used is Eviews 5.0.
*** indicates significance at 5% and ** at 10%.

As shown in Table 13.7, these factors all have the expected, positive effect on the TFP. Especially, the coefficients of the growth rate of the expenditure for extension services and knowledge diffusion and the ratio of the area of converted sloping cropland to total area of cropland are highly significant and large in magnitude. These results confirm that cropland reduction should not be inevitably associated with a TFP reduction, and that the extension services and diffusion of technical knowledge have played a vital role in promoting TFP growth. As such, these efforts should be strengthened in the future. It should also be pointed out that while the majority of the surveyed households experienced a TFP gain, a number of them suffered a TFP decline. This implies that the local government agencies must pursue more effective targeting in not only land conversion, but also production assistance and poverty reduction (Chen, 2006).

At the same time, a reconfiguration of the production technology and thus its frontier should have been expected, given the huge reduction of cropland and native goats, and the major restructuring of farming and animal husbandry. In this context, the measured decline in technological change is understandable; this longer-term issue requires coordinated and persistent efforts to tackle. As to the slight decline of scale efficiency, it is reasonable given that farmland was reduced, open grassing of large flocks was banned, and enclosed pen-feeding limited the scale of animal production. Also, the emergence of a dynamic energy industry absorbed a certain amount of rural labor, restraining the scale of agricultural production (Wuqi Statistics Bureau, 2003).

Finally, it should be made clear that because this study is based on data from Wuqi and for the period of 1998–2004, caution needs to be taken in applying the findings to another area or a longer period of time. In the future, this type of work should be pursued for other regions and/or periods, and the other factors affecting TFP, such as changes in natural and social conditions, should be considered in order to measure the program performance more thoroughly.

Acknowledgements This study is sponsored by China's National "11th Five-Year Plan" Science and Technology Support Project (2006 BAD03A0308), the International Research Center of Sediment project of "A Study of Wuqi County's Sustainable Development" (2005-01-10), and the

Shaanxi Provincial Forestry Special project "Ecological Forest Policy Research." The authors are grateful for Prof. Runsheng Yin's help in editing this chapter.

References

Bennett, M. T. (2008). China's sloping land conversion program: institutional innovation or business as usual? *Ecological Economics, 65*, 700–712.

Chen, W. P. (2006). China's agricultural productivity growth, technological progress, and efficiency changes between 1990 and 2003. *China Rural Survey, 1*, 18–23.

Coelli, T. J., Rao, D. S. P., & Battese, G. E. (1998). *An introduction to efficiency and productivity analysis*. Norwell, MA: Kluwer Academic Publishers.

Färe, R., & Grosskopf, S. (1992). Malmquist productivity indexes and Fisher ideal indexes. *The Economic Journal, 102*, 158–160.

Färe, R., & Primont, D. (1995). *Multi-output production and duality: Theory and application*. Boston, MA: Kluwer Academic Publishers.

Feng. Z, Yang, Y., Zhang, Y., Zhang, P., & Li, Y. (2005). Grain-for-Green policy and its impacts on grain supply in west China. *Land Use Policy, 22*(3), 301–312.

Gopinath, M., & Roe, T. L. (1997). Efficiency and technical progress: Sources of convergence in the Spanish regions. *Applied Economics, 32*(4), 467–478.

Huang, J. K., & Rozelle, S. (1996). Technological change: Rediscovering the engine of productivity growth in China's rural economy. *Journal of Development Economics, 49*(2), 337–367.

Lin, Y. F. (1992). Rural reforms and agricultural growth in China. *American Economic Review, 82*, 34–51.

Liu, C. (2003) An investigation of Jinzhai County's farming efficiency and poverty alleviation. *Quantitative Economic and Technological Research, 12*, 102–106.

Liu, C., & Zhang, W. (2006). Impacts of conversion of farmland on household income: Evidence from implementing the Desertification Combating Program in the Vicinity of Beijing and Tianjin. *China Economics Quarterly, 23*, 273–290 (in Chinese).

Lissitsa, A., & Odening, M. (2005). Efficiency and total factor productivity in Ukrainian agriculture in transition. *Agricultural Economics, 32*(3), 311–325.

Managi, S., Opaluch, J. J., Jin, D., and Grigalunas, T. A. (2004). Technological change and depletion in offshore oil and gas. *Journal of Environmental Economics and Management, 47*: 388–409.

State Forestry Administration (SFA). (2003). *A monitoring and assessment report on the socio-economic impacts of China's key forestry programs*. Beijing, China: China Forestry Press.

Uchida, E., Xu, J. T., Xu, Z. G., & Rozelle, S. (2005). Grain for Green: Cost – effectiveness and sustainability of China's conservation set aside program. *Land Economics, 81*, 247–264.

Wuqi SLCP Office. (2007). *A summary of Wuqi's implementation of the sloping land conversion program*.

Wuqi Statistics Bureau. (2003) *Wuqi's 2002 Statistics Yearbook*.

Xu, J. T., Tao, R., & Xu, Z. G. (2004). Sloping land conversion: Cost-effectiveness, structural adjustment, and economic sustainability. *China Economics Quarterly, 4*(1), 139–162 (in Chinese).

Xu, J. T., Yin, R. S., Li, Z., & Liu, C. (2006) China's ecological rehabilitation: Unprecedented efforts, dramatic impacts and requisite policies. *Ecological Economics, 57*, 595–607.

Xu, Q. (2005). An empirical analysis of agricultural efficiency variation in Zhejiang. *Ningbo University Journal, (Technology Edition) 2*, 16–20 (in Chinese).

Xu, W., Yin, Y. Y., & Zhou, S. (2007). Social and economic impact of carbon sequestration and land use change on peasant households in rural China: A case study of Liping, Guizhou Province. *Journal of Environmental Management, 85*(3), 736–745.

Yi, G., Fan, G., & Li, Y. (2003). A theoretical analysis of economic growth and total factor productivity in China. *Economic Research Journal, 8*, 13–19 (in Chinese).

Yin, R. S., Xu, J. T., Li, Z., & Liu, C. (2005). China's ecological rehabilitation: The unprecedented efforts and dramatic impacts of reforestation and slope protection in western China. *China Environment Series, 6*, 17–32.

Yin, R. S., & Yin, G. P. (2009). China's ecological restoration programs: Initiation, implementation, and challenges. *Environmental Management,* (submitted).

Zhang, J., & Shi, S. H. (2003). A change of total factor productivity for the Chinese economy: 1952–1998. *World Economic Forum, 2*, 17–24 (in Chinese).

Zhang, S. Q., Swanson, T., & Kontoleon A. (2005). *Impacts of the compensation policies in reforestation programs.* A Report to the Environment and Poverty Program of China Council for International Cooperation in Environment and Development.

Chapter 14
Measuring the Aggregate Socioeconomic Impacts of China's Natural Forest Protection Program

Yueqin Shen, Xianchun Liao, and Runsheng Yin

Abstract China has been implementing one of the world's largest ecological rehabilitation projects, the Natural Forest Protection Program (NFPP), to improve its fragile and precarious environmental conditions. This chapter measures the socioeconomic impacts of the NFPP using input–output (I–O) models. We find that the NFPP will expand the annual output of the forest sectors by 5.8 billion Yuan and the whole economy by 8.9 billion Yuan by 2010. Employment will increase by 0.84 million in the forest sectors and by 0.93 million in the whole economy. Associated with the enormous expansion of forest protection and management are huge contributions to mitigating water runoff, soil erosion, flooding, and biodiversity loss. The investments and adjustments are thus worthwhile, if the program is properly implemented. The challenges are to transform loggers into tree planters and forest managers and to ensure that the financial and institutional commitments by the local and national governments will be materialized.

Keywords Natural Forest Protection Program · Input–output analysis · Socioeconomic impact · Environmental protection

14.1 Introduction

China has been implementing one of the world's largest ecological rehabilitation projects, the Natural Forest Protection Program (NFPP), to improve its fragile and precarious ecosystem conditions. Zhang et al. (2000) reported on the NFPP and discussed policy measures for its implementation. Loucks et al. (2001) argued that the NFPP could strengthen the pandas' future in China's forests by enhancing protection and restoration of corridors among remaining forest fragments and increasing habitat preservation. Zhang et al. (1997); Zhao and Shao (2002) noted the logging

Y. Shen (✉)
School of Economics and Management, Zhejiang Forestry University, Lin'an 311300, P.R. China
e-mail: shenyueqin@zjfc.edu.cn

R. Yin (ed.), *An Integrated Assessment of China's Ecological Restoration Programs*,
DOI 10.1007/978-90-481-2655-2_14, © Springer Science+Business Media B.V. 2009

restrictions induced by the NFPP and their potential economic and environmental impacts. While the NFPP has drawn broad attention, little has been done to measure its potential environmental and economic impacts. The goal of this chapter is to tackle this important issue using an input–output (I–O) analysis based on the recent Chinese national statistics.

Population growth, economic development, and policy failures have resulted in severe environmental problems in China, such as loss of biodiversity, desertification, and soil erosion (World Bank, 1994; Fullen & Mitchell, 1994; Zhang et al., 2000). While the country has made efforts to combat these problems (SFA, 2000), they have not been very effective, and the macro ecological conditions have worsened. In particular, the 1998 floods along the Yangtze River and waterways in the northeast devastated large parts of China, leading to the loss of more than 3,000 human lives and US \$12 billion in property damage and output reduction (Lu et al., 2002; Liang, 1998). It has been accepted that the floods were caused mainly by deforestation and farming on steep slopes (Lu et al., 2002; Xu, Yin, Li, & Liu, 2006). In response, the government has, among other things, initiated the NFPP to protect and expand its forest resources throughout this decade.[1]

While the primary objective of the NFPP is to protect the existing forests, it also aims to expand their coverage through natural regeneration and artificial planting in order to mitigate the occurrence and influence of natural disasters (Zhang et al., 2000). To achieve these objectives, bans of commercial logging are imposed in the southwest, harvests are substantially reduced in the northeast and other regions, and forestation and vegetation activities are carried out.

According to the government plan, the protection and management of over 95 million ha of forestland will be greatly strengthened.[2] In addition, 0.5 million ha of bare or degraded land will be afforested every year, and 1.4 million ha of mountains and hillsides will be closed for natural regeneration. Forest cover in the targeted areas will increase from the current 17.5–26.8% by 2010 (SFA, 2000). Table 14.1 summarizes the proposed activities of forest protection, management, and expansion during 1998–2010. The Chinese government hopes that the NFPP and other initiatives will not only greatly improve the domestic ecological conditions but also significantly contribute to the regional and even global environmental protection.

Altogether, the NFPP covers 17 provinces and autonomous regions, including 414 counties or forest bureaus in the upper reaches of the Yangtze River (from the Three Gorges Dam upward) and 358 counties or forest bureaus in the middle and upper reaches of the Yellow River (from the Xiaolangdi Dam upward). In addition, it encompasses 84 state forest bureaus, twelve provincial forest enterprises, and one county in the northeast; four provincial forest enterprises and seven county-level forest farms in Hainan; and two state forest bureaus, 25 counties, and four county-level forest farms in Xinjiang (SFA, 2000).

---

[1] The Sloping Land Conversion, or "Grain for Green," Program is another new initiative of ecological rehabilitation. For more details, see Forest and Grassland Taskforce (2003) and Xu et al. (2006).

[2] China has a total forestland of 155 million ha (Xu et al., 2006).

Table 14.1 Forest Regeneration and Tree Planting Under the NFPP

Area planted or managed	Unit (ha)	1998	1999	2000	2005	2010
Total afforestation	1000	290	478	775	549	549
Artificial regeneration	1000	289	441	303	344	344
Aerial seeding by airplane	1000	1	37	152	205	205
Human-facilitated natural regeneration	1000	139	378	259	259	259
Mountain closure	1000	2,433	2,207	816	1,463	1,463
Stand tending	1000	882	1,679	1,281	1,281	1,281

Note: Data sources include the China Forestry Development Report (SFA, 2000–2002) and the China Forestry Statistics Yearbook (1998–1999).

Given the lack of relevant data, this chapter does not attempt to assess the environmental impacts of the NFPP. Rather, it will focus on the NFPP's socioeconomic impacts on various forest sectors as well as the whole economy. By socioeconomic impacts we mean the concomitant changes in employment, wages and salaries, total output, and value added. In this context, input–output (I–O) analysis is an appropriate technique because it emphasizes inter-sectoral linkages of the economy and can be used to quantify the impacts of either changes of final demands on various economic sectors, or changes of the output in one sector on the other sectors and the whole economy. Luckily, we have obtained China's national I–O table for 1997, which is a valuable data source to support our analysis. Of the 124 sectors included in the national I–O table, five are identified as forest sectors: forest management, logging and hauling, sawmilling and panel production, furniture and solid wood products, and pulp and paper making.

The socioeconomic impacts of the NFPP on a specific sector or the whole economy come from two sources – the increased investments in forest protection and management, and the logging bans and harvest reductions. Notably, these changes have distinct features. First, impacts induced by the former should be positive, whereas impacts induced by the latter should be negative. Second, the former are changes in final demand, while the latter are changes in the output of one sector. The objective of this study is to quantify the socioeconomic impacts caused by these different changes. We believe that this study is important and timely for better understanding the potential ramifications of the NFPP and its more effective implementation.

Researchers have extensively applied I–O analysis to forestry (Davis et al., 2000), but few have simultaneously considered the impacts of both changes in final demands and changes in the outputs of certain sectors. For instance, Munn (1998) and Wu (2002) used I–O analysis to assess the importance of the forest products industry in Mississippi and Texas to the state economies. However, they focused only on the economic impacts of final demand changes, with little consideration of the impacts of exogenous output changes of one sector on the other sectors. In the context of the NFPP, both changes in final demands and outputs in the logging and hauling sector are relevant. As to the recent case studies done by Chinese scholars (e.g., Chen, Xiang, Liu, & Mu, 2001; Liu, 2002; Shen, 2000; Sheng, 2002) in examining the impacts of the NFPP, they are descriptive and preliminary, and little in-depth analysis was conducted. Also, limited attention was given to the linkages of the economy.

The chapter is organized as follows. First, we articulate the specific impacts of the NFPP in section two to motivate the development of I–O models in section three. In section four, a base case (without the NFPP) is first presented, followed by an analysis of the effects of the policy (with the NFPP). The two scenarios are then combined to determine the impacts of the NFPP on the forest sectors and the whole economy. Finally, conclusions are summarized and suggestions for future work are made in section five.

14.2 Impacts of the NFPP

As noted, the socioeconomic impacts of the NFPP may be induced by the logging bans and harvest reductions, or by the increased investments in forest protection and management. Before we use I–O models to quantify them, it is necessary to define these specific impacts and their sources.

First, the NFPP has significantly reduced timber supply. Roundwood production from natural forests decreased from 32.1 million m^3 in 1997 to 29.3 million m^3 in 1998 and further to 22.8 million m^3 in 1999. Between 2003 and 2010, roundwood production from natural forests will be maintained at 12.1 million m^3 (SFA, 2000), implying that the NFPP will cause a reduction of 20 million m^3/yr in roundwood production throughout this decade. Since the NFPP has mainly reduced the supply of large- and medium-diameter logs (>14 cm), many sawmills have been facing operation difficulty or even shut down. Furthermore, the enlarged gap between timber demand and supply has caused timber prices to rise. For example, it is reported that the log prices in Beijing area increased by 20–30% in 1998 (Studley, 1999). Partly to fill this gap and partly to meet the growing demand, China has turned increasingly to the international market. Figure 14.1 shows that the imports of timber products have increased sharply, following the implementation of the NFPP.[3]

Because of the logging bans and harvest reductions, the logging, hauling, and processing equipment and facilities owned by the state, worth almost 15 billion Yuan (US\$ 1.8 billion), have become obsolete and thus been abandoned (Li, 2001). Also, an additional one billion Yuan/yr (US\$ 120 million) of interest payment for bank loans – an obligation of state forest enterprises – has been accumulated, waiting for write-off by the central government (SFA, 2000; Xu et al., 2006). Meanwhile, the NFPP has laid off a large number of workers in the logging, hauling, and processing sectors, with 0.5 million employees transferred to the forest protection and management activities (SFA, 2000). As a result, the payments for wages and salaries in these sectors have dwindled.

---

[3] As pointed out by a reviewer, in addition to alleviating the pressures on domestic forests, increased timber imports enable the maintenance of a larger share of the wood products processing facilities, which will further swing the balance towards positive net effects of the NFPP.

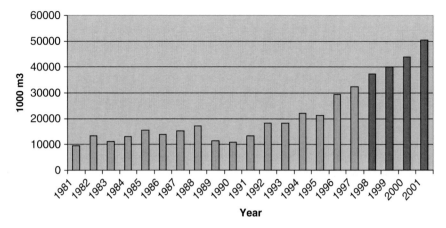

Fig. 14.1 China's Timber Imports Since 1981. (Notes: 1. Data are from China Customs Statistics (1981–2001). 2. Volumes of Different Products are Converted to Roundwood Equivalence)

In comparison, forest protection and management activities have been expanded substantially because of the government investments. Table 14.2 shows that the total expenses in these activities would reach 121 billion Yuan (US$ 14.6 billion) by 2010 – more than 10 billion Yuan a year. In accordance with these structural adjustments, more permanent as well as temporary employees have been added, and the total payments for wages and salaries in forest protection and management have increased tremendously.

Similarly, the NFPP has affected the other sectors of the economy. On the one hand, logging bans and harvest reductions have reduced the total output, employment, wages and salaries, and value added; on the other, these socioeconomic

Table 14.2 Governmental Financial Support for Implementing the NFPP Since 1998 (Unit: 1000 yuan)

Year	Total expenditure	Infrastructure construction	Equipment and facilities for fire control, etc.	Afforestation and forest management	Compensation to workers
1998	4,341,260	62,070	57,120	2,158,420	2,063,650
1999	7,605,340	49,360	28,450	4,014,440	3,513,090
2000	9,493,190	61,612*	35,512*	5,010,932*	4,385,133*
2001–2010*	100,000,000	649,018*	374,079*	52,784,491*	46,192,412*

Notes:
1. Data sources include the China Forestry Development Report (SFA, 2000 2002) and the China Forestry Statistics Yearbook (1998–1999).
2. An outlay of 100 billion Yuan (US$ 12 billion) has been planned for the period of 2001–2010 (Zhao & Shao, 2002; Xu et al., 2006).
3. Afforestation and forest management include seedling production, mountain closure, stand tending, artificial and natural generation, and pest control.
* indicates that data are estimated based on the allocation made in 1999.

measures have been expanded due to the increased governmental investments. For instance, in local communities near the natural forests, farmers, who had attached themselves to the logging, hauling, and processing activities through direct employment or indirect services, have been hit hard by the NFPP. Their losses were estimated at 2.3 billion Yuan (US\$ 277 million) of sales revenue a year (Yu, Xie, Li, & Chen, 2002), and their net income was reduced by 18% (Chen et al., 2001). But farmers have also gained from the new job opportunities of tree planting (and even forest protection) due to the combined effect of its seasonality and large workload. As another example, it was estimated that the NFPP reduced the value-added and employment of railway freight by 223 million Yuan and 14,000 workers, respectively, in 2000 (SFA, 2000).

In short, the forest sectors have extensive interactions with other sectors of the economy. The implementation of the NFPP results in not only intra-sectoral but also inter-sectoral impacts due to their interdependency, and these impacts can be categorized as direct, indirect, and induced effects. The NFPP's direct effects refer to the changes in production, employment, wages and salaries, and value-added caused to the forest sectors. The NFPP's indirect effects refer to the changes of economic activities in other sectors, resulting from the forest sectors' altered purchase of their goods and services. The induced effects of the NFPP come from changes in consumption of goods and services by the employees in the forest sectors. The total socioeconomic impacts of the NFPP on the forest sectors thus include all these effects. Here, the overarching questions are: How will the new forestry program affect the forest sectors and other sectors? What will be the positive impacts induced by the government investments and the negative impacts induced by the logging bans and harvest reductions? By how much will the former be larger than the latter? We address these questions below.

14.3 I–O Models and Data

Socioeconomic impacts can be measured based on the transaction relationships embedded in an I–O table. Referring to Table 14.3, the transactions in an I–O table can be divided into sub-matrices X, Y, and Z. Let $X = \{x_{ij}\}$, where x_{ij} is the amount of sales from sector i to sector j or the amount of purchases from sector j to sector i $(i, j = 1, 2, \ldots, n)$; $Y = \{y_i\}$, where y_i is the final demand from sector i; and

Table 14.3 A Stylistic Input–Output Table

Producing sector (i)	Purchasing sector (j)					Total sales
	(1)	(2)	(3)	(4)	(5)	
(1)	x_{11}	x_{12}	x_{13}	x_{14}	y_1	z_1
(2)	x_{21}	x_{22}	x_{23}	x_{24}	y_2	z_2
(3)	x_{31}	x_{32}	x_{33}	x_{34}	y_3	z_3
(4)	x_{41}	x_{42}	x_{43}	x_{44}	y_4	z_4
Total purchases	z_1	z_2	z_3	z_4		

$Z = \{z_j\}$, where z_j is the total purchase of sector j. By definition, the total output of each sector is equal to the intermediate demand from other industries and the final demand; likewise, the total input of each sector is equal to the intermediate inputs from other industries and the final payment. So, the sales and purchases of a given sector are equal; that is, $z_i = z_j$ for $i = j$.

Further, let $A = \{a_{ij}\}$, where $a_{ij} = x_{ij}/z_j$ is the portion of total purchases by sector j from sector i. Then, from $X = A \times Z$ and $X + Y = Z$ we know that $Y = (I - A) \times Z$, where I is an $n \times n$ identity matrix. Thus,

$$Z = (I - A)^{-1} \times Y \qquad (14.1)$$

The above equation links the level of total economic activity Z to final demand Y by the multiplier matrix $(I - A)^{-1}$ derived from the I–O table. Note that the column sums of the multiplier matrix represent the total effects on the outputs of different sectors of the economy given a unit change in final demand, which are called the output multipliers.

The multiplier matrix reveals the linkage between sectors in the economy and presents two types of multipliers: type I and type II. Type I multipliers sums together direct and indirect impacts while type II multipliers also include induced effects. In this chapter, type I multipliers are adopted. This is because treating household consumption as endogenous will entail forecasting its future values with its identified determinants, which is beyond what our data can accommodate at this point of time. Also, in a developing economy like China's, where a substantial part of the labor force does not participate effectively in the production process, it may not be easy to determine the relation between population size and total output (Hubacek & Sun, 1999). In any case, we expect that the induced effect should be only a small portion of the total effect.

14.3.1 Response to Final Demand Changes

Depending on the sources of exogenous changes, the policy impacts on certain sectors and/or the whole economy may be assessed in different ways. If the vector of final demands changes by ΔY, Equation (14.1) can be written as

$$\Delta Z = (I - A)^{-1} \times \Delta Y \qquad (14.2)$$

thus, a vector of total output changes (ΔZ) can be computed directly. In the current context, changes in final demands are changes in the government investments in forest protection and management.

Based on the estimated multipliers, it is possible to derive employment changes as well. The formula for the employment change in sector j is

$$\Delta \mu_e = \sum_i (e_i/q_i) * INV_{ij} \Delta Y \qquad (14.3)$$

where $\Delta\mu_e$ represents employment change due to the final demand change, e_i is the number of employees in sector i, q_i denotes the total output in sector i, and INV_{ij} is an element of multiplier matrix $(I - A)^{-1}$ (Schaffer, 1999).

Similarly, the wages and salaries change can be obtained from the formula

$$\Delta\mu_h = \sum_i (h_i/q_i)^* INV_{ij} \Delta Y \tag{14.4}$$

where $\Delta\mu_h$ represents wages and salaries change driven by the final demand change, and h_i is the household income of sector i. INV_{ij} and q_i are defined the same as above.

Likewise, the value-added change can be calculated from

$$\Delta\mu_v = \sum_i (v_i/q_i)^* INV_{ij} \Delta Y \tag{14.5}$$

where $\Delta\mu_v$ represents value-added change derived from the final demand change, and v_i is the value-added of sector i.

14.3.2 Response to Output Changes

In many I–O models, only the final demand is considered exogenous. But a mixed type of I–O model may be employed, in which final demands for some sectors and gross outputs for the remaining sectors are specified exogenously (Miller & Blair, 1985). In our case, the output of the logging and hauling sector is determined exogenously. If the output of sector h, $\overline{Z}_h$, is also exogenously determined, Equation (14.1) can be modified as follows:

$$\begin{bmatrix} Z_1 \\ . \\ . \\ Z_{h-1} \\ Z_{h+1} \\ . \\ . \\ Z_n \end{bmatrix} = (I - A^*)^{-1} \begin{bmatrix} Y_1 + a_{1h}\overline{Z}_h \\ . \\ . \\ Y_{h-1} + a_{h-1,h}\overline{Z}_h \\ Y_{h+1} + a_{h+1,h}\overline{Z}_h \\ . \\ . \\ Y_n + a_{nh}\overline{Z}_h \end{bmatrix} \tag{14.6}$$

where A^* is a new $[(n{-}1) \times (n{-}1)]$ input coefficient matrix that differs from A above. Equation (14.6) can be further modified as:

$$
\begin{bmatrix} \Delta Z_1 \\ . \\ . \\ . \\ \Delta Z_{h-1} \\ \Delta Z_{h+1} \\ . \\ . \\ \Delta Z_n \end{bmatrix} = (I - A^*)^{-1} \begin{bmatrix} a_{1h}\Delta \overline{Z}_h \\ . \\ . \\ . \\ a_{h-1,h}\Delta \overline{Z}_h \\ a_{h+1,h}\Delta \overline{Z}_h \\ . \\ . \\ a_{nh}\Delta \overline{Z}_k \end{bmatrix}
\tag{14.7}
$$

Combining Equations (14.3), (14.4) and (14.5) with Equation (14.7), we can examine the negative impacts of the roundwood output decline caused by the logging bans and harvest reductions on the other forest sectors and the overall economy as well. For example, the impacts of the logging bans and harvest reductions on employment in sector h can be measured by:

$$
\Delta \mu_e^* = \sum_i (e_i/q_i)^* INV_{ij}^{**} \begin{bmatrix} a_{1h}\Delta \overline{Z}_h \\ . \\ . \\ . \\ a_{h-1,h}\Delta \overline{Z}_h \\ a_{h+1,h}\Delta \overline{Z}_h \\ . \\ . \\ a_{nh}\Delta \overline{Z}_h \end{bmatrix}
\tag{14.8}
$$

where $\Delta \mu_e^*$ represents employment changes due to the output change in the logging and hauling sector, e_h is the number of employees, q_h denotes the total output, and INV_{ij}^* is an element of the multiplier matrix $(I - A^*)^{-1}$.

14.3.3 I–O Data

We used China's 1997 national I–O table for our analysis (State Statistics Bureau, or SSB, 1998). Since 1981, the SSB has compiled eight national I–O tables. In collaboration with Hong Kong Chinese University, the SSB has recently converted these tables into a uniform format in constant prices. Of the 124 sectors in the national I–O tables, five are identified as forest sectors: forest management, logging and hauling, saw-milling and panel production, furniture and solid wood products, and pulp and paper making. Three other sectors closely related to the forest sectors – coal mining, construction, and railway freight – are also identified.

To focus on the sectors of our primary interest and to reduce the involved workload, an aggregation of other sectors was made. We combined the four agricultural sectors – cropping, fisheries, livestock, and other agricultural production – into one, and the three printing and cultural sectors – printing and recording media, cultural goods, and toy production and recreation – into one. Likewise, we combined the ten chemical production-related sectors into one, the six transportation equipment-related sectors into one, the sixty other sectors in secondary industry into one, the ten production sectors in the tertiary industry into one, and the 23 service sectors in the tertiary industry into one. As a result, the final table for analysis contained 15 sectors.

The 1997 data reflect socioeconomic activities just one year before the initiation of the NFPP, and thus matched our base case nicely. All of the estimated monetary values are in 1997 Chinese Yuan. Employment data were taken from China's 1997 Statistics Yearbook and Forestry Yearbook. Other data for the forest sectors were obtained from the China Forestry Development Reports (SFA, 2000–2002).

Before proceeding to the presentation of our results, a brief discussion of some data issues are warranted. First, someone may question the way we aggregated the 124 economic sectors into 15 for analysis. Our view is that to make our analysis as well as presentation practical, it is necessary to cut down the number of sectors we deal with. There are two approaches to do so. One is to aggregate the 124 sectors first and then create the inverse matrix of the aggregated sectors and do the necessary calculations. The other is to create the inverse matrix of the 124 sectors and do the necessary calculations first and then aggregate the calculated results for presentation. Conceptually, the latter is the better approach. But the former was used for data and computational reasons in this chapter.

In addition, China's statistics in general and its GDP in particular are suspected of inconsistency and inaccuracy, due to the changing procedures of data gathering and local officials' "obsession" with GDP growth rates – the leading criterion for evaluating cadre performance (Rawski, 2001). While we acknowledge these problems and our concern with their influence on data and thus analytic reliability, we think that their potential effect on our study is small. Primarily, given the way the multiplier matrix $(I - A)^{-1}$ is derived, these problems are pretty much offset. Also, even if the base-year (1997) economic activities had been overstated, say, by 2%, its direct impact on our aggregate assessment would be very limited.

Moreover, one may argue that the figures of the initial NFPP impacts, reported by the State Forestry Administration, could be an overestimation of its positive effects and an underestimation of its negative effects. After extensive field visits and discussions with many scholars, our finding is that, while it is quite likely that the official estimates are incomplete, it is less likely that they have been intentionally manipulated. In any case, caution is warranted in interpreting our results and drawing policy implications.

14.4 Results

In this section, we first show the direct and total effects of the forest sectors on the economy in 1997 as a base case, which is meant to provide a picture of what roles the forest sectors then played. Next, we present the policy scenario with the NFPP, including its positive and negative impacts on the forest sectors, and on the whole economy in terms of total output, employment, wages and salaries, and value-added.

14.4.1 The Base Case

China's forest sectors are important to the whole economy (Table 14.4). In 1997, the country produced 63.9 million m^3 of roundwood, 20.1 million m^3 of lumber, 16.5 million m^3 of wood-based panels, and 44.8 million tons of pulp and paper products (including products made of non-wood fibers). In addition, the forest sectors produced such value-added products as veneers, woodchips, and furniture products. Together, these sectors employed 17.1 million workers, and paid 133.0 billion Yuan in wages and salaries. In addition, they generated outputs worth 580.5 billion Yuan, of which value-added amounted to 213.4 billion Yuan.

Table 14.4 also shows that in terms of employment, the forest management sector was ranked first in 1997, accounting for 71.1%. The paper and paperboard sector was ranked second (13.0%), and the furniture-making sector was ranked the last (3.2%). As to total output, the paper and paperboard sector was the largest (42.0%), the sector of furniture making and other wood products ranked second (23.6%), whereas the logging and hauling sector was the smallest (5.2%). For value-added, the paper and paperboard sector was the largest (33.6%), the forest management sector was ranked second (28.4%), while the logging and hauling sector was the smallest (8.7%).

Table 14.5 summarizes the total effects for the forest sectors. The estimated output effects of the forest sector were 1.55 trillion Yuan in 1997. The output share of the forest sectors accounted for 8% of gross domestic output, of which the

Table 14.4 The Direct Effects of the Forest Sectors on China's Economy in 1997

Sector	Number of employees	Total industry output (1000 yuan)	Wages and salaries (1000 yuan)	Value-added (1000 yuan)
Forest management	12,152,000	82,587,000	53,458,412	60,538,329
Logging and hauling	962,858	30,015,155	8,682,428	18,639,838
Sawmilling and panel production	1,215,919	87,119,576	12,112,873	26,576,558
Production of furniture and woodworks	542,720	136,998,635	18,024,779	36,037,235
Paper and paperboard production	2,227,609	243,754,236	40,732,149	71,594,631
Total	17,101,105	580,474,602	133,010,641	213,386,591

Notes:
1. Employment data were taken from the China Statistics Yearbook (1997) and the China Forestry Statistics Yearbook (1997). Based on available information, the authors estimated employees in the forest sectors. Other data were from the national input–output table (State Statistics Bureau, 1997).
2. All monetary values are in 1997 yuan.

Table 14.5 Total Economic Impacts of the Forest Sectors on China's Economy in 1997

Sector	Number of employees	Total industry output (1000 yuan)	Wages and salaries (1000 yuan)	Value-added (1000 yuan)
Forest management	14,385,868	134,534,223	66,493,911	72,928,506
Logging and hauling	2,145,001	56,303,064	15,513,229	13,981,045
Sawmilling and Panel production	6,270,840	237,313,725	45,052,224	39,706,736
Production of furniture and woodworks	7,688,145	411,954,895	69,320,103	68,522,741
Paper and paperboard production	15,365,932	711,274,861	133,028,394	138,678,470
Total	45,855,786	1,551,380,768	329,407,861	333,817,499

Notes:

1. Total economic impacts were estimated by multiplying the multipliers with the direct effects of the forest sectors.

2. All monetary values are in 1997 yuan.

value-added share of the forest sectors was 4% of the national GDP.[4] Also, they provided 45.9 million jobs, and paid 329.4 billion Yuan in wages and salaries.

14.4.2 The Policy Scenario

The impacts of the NFPP on the forest sectors and on the whole economy are considered over the period 1998–2010.[5] The increased government investments are used to: (1) support construction projects required for forest management and fire control, such as house, road, and fire tower building; (2) purchase equipment and vehicles for fire control and transportation, and (3) increase seed and seedling production, stand tending, artificial regeneration, pest control, and other activities.

In contrast, the logging bans and harvest reductions have resulted in decreases in roundwood production, employee lay-offs, wages and salaries cuts, and a decline

---

[4]Despite the bottleneck nature of the forest sectors to the Chinese economy, the shares of industry output and GDP are much higher compared to those, say, for the United States. In addition to China's stage of economic development, a main reason is that the country takes the gross value of the standing forests into account.

[5]Since we intended to focus on the socioeconomic impacts of the NFPP in this study, we decided not to consider questions related to changes in timber production and forest management outside of the NFPP coverage and the increased imports of forest products. Certainly, they can be addressed in a similar manner.

of value-added in the logging and hauling sector. The reduced log production has in turn affected the production in the other forest sectors. Because timber production from natural forests would be maintained at a constant level after 2003 (SFA, 2000), no more changes should thus be expected in annual logging and hauling ($\Delta \bar{Z}_h = 0$).

Tables 14.6, 14.7, 14.8, and 14.9 summarize the impacts of the NFPP on the forest sectors. Compared to the 1997 base case, the total output of the forest sectors was reduced by 1.3 billion Yuan in 1998 due to the logging bans and harvest reductions. However, the government investments led to an output increase in the forest sectors by 2.4 billion Yuan in the same year. The net output in the forest sectors thus gained 1.1 billion Yuan. The annual output of the forest sectors will expand by 5.8 billion Yuan by 2010. In addition, there were 0.04 million laid-off employees, while the government investments added 0.34 million jobs in the forest sectors in 1998. The net increase of employment in the forest sectors was 0.3 million. While the logging and hauling sector has suffered substantial job reductions during the past several years, the employment of the forest management sector has been expanding. Notably, these results are very close to the figures reported in the China Forestry Statistics Yearbook (1998, 1999). The total employment of the forest sectors is projected to increase by 0.84 million by 2010.

The reduced log production caused a loss of value-added in the forest sectors by 0.8 billion Yuan in 1998, but the increased investments in forest management resulted in a value-added gain by 1.7 billion Yuan. The annual value-added in the forest sectors will increase by 4.2 billion Yuan in 2010, of which wages and salaries will account for 3.7 billion Yuan. In short, the implementation of the NFPP will greatly benefit the forest sectors from the increased governmental investments, although the benefits come with a significant cost to the logging and hauling sector.

Likewise, the NFPP has both positive and negative effects on the whole economy (Tables 14.6, 14.7, 14.8, and 14.9). Compared to the 1997 base case, the annual industry output of the whole economy will expand by 8.9 billion Yuan, and the annual employment will increase by 0.93 million by 2010. Also, the NFPP will augment the annual value-added by 5.4 billion Yuan, of which wages and salaries will increase by 4.3 billion Yuan. As a result, implementing the NFPP will benefit the whole economy as well. For instance, the annual output of the agricultural sector will expand by 0.3 billion Yuan, and its annual employment will increase by 39,000 by 2010.

14.5 Conclusions and Discussion

Using an I–O analysis, this chapter has assessed the socioeconomic impacts of the NFPP on the forest sectors and the whole Chinese economy. The advantage of this approach is its ability to measure both intra-sectoral and inter-sectoral linkages, and to examine the economic responses to changes in the final demand and/or the output of a certain sector. To sum up, the NFPP will expand the annual output of the

Table 14.6 Output Changes in the Forest Sectors and the Whole Economy Induced by the NFPP (Unit: 1000 yuan)

Sector		1998	1999	2000	2005	2010
Forest	Increase	2,329,344	4,330,925	5,405,977	5,694,585	5,694,585
Management	Decrease	155,710	367,775	504,211	0	0
	Net change	2,173,634	3,963,150	4,901,766	5,694,585	5,694,585
Logging and	Increase	6,765	11,600	14,480	15,253	15,253
hauling	Decrease	1,081,212	2,553,741	3,501,122	0	0
	Net change	−1,074,447	−2,542,141	−3,486,642	15,253	15,253
Sawmilling and	Increase	7,502	11,973	14,945	15,743	15,743
panel production	Decrease	3,976	9,391	12,875	0	0
	Net change	3,526	2,582	2,070	15,743	15,743
Production of	Increase	4,915	8,047	10,045	10,581	10,581
furniture and	Decrease	2,386	5,637	7,728	0	0
woodworks	Net change	2,529	2,410	2,317	10,581	10,581
Paper and	Increase	22,687	37,352	46,624	49,113	49,113
paperboard	Decrease	10,356	24,460	33,533	0	0
production	Net change	12,331	12,892	13,091	49,113	49,113
Subtotal net change		1,117,573	1,438,893	1,432,602	5,785,275	5,785,275
Other sectors	Increase	1,495,083	2,369,734	2,957,965	3,115,882	3,115,882
	Decrease	740,516	1,749,044	2,397,900	0	0
	Net change	754,567	620,690	560,065	3,115,882	3,115,882
Total net change		1,872,140	2,059,583	1,992,667	8,901,157	8,901,157

Notes:

1. The increase of industry output is induced by the government investments in forest protection and management, whereas the decrease is induced by the logging bans and harvest reductions.

2. See discussion on pp. 4–5 for logging bans and harvest restrictions over time; and see Table 14.2 for investments in forest protection, management, and expansion.

3. The results are reported in annual changes for comparing with the base case.

Table 14.7 Annual Changes of Employees in the Forest Sectors Induced by the NFPP

Sector		1998	1999	2000	2005	2010
Forest management	Increase	342,744	637,331	795,551	838,061	838,061
	Decrease	22,945	54,194	74,299	0	0
	Net change	319,799	583,137	721,252	838,061	838,061
Logging and hauling	Increase	217	378	474	502	502
	Decrease	15,909	375,762	515,161	0	0
	Net change	−15,692	−375,384	−514,687	502	502
Sawmilling and panel production	Increase	105	171	214	228	228
	Decrease	56	131	180	0	0
	Net change	49	40	34	228	228
Production of furniture and woodworks	Increase	19	33	42	45	45
	Decrease	9	22	31	0	0
	Net change	10	11	11	45	45
Paper and paperboard production	Increase	207	362	457	492	492
	Decrease	95	224	307	0	0
	Net change	112	138	150	492	492
Subtotal net change		304,278	207,942	206,760	839,328	839,328
Other sectors	Increase	40,544	69,843	88,075	94,718	94,718
	Decrease	18,244	43,090	59,075	0	0
	Net change	22,300	26,753	29,000	94,718	94,718
Total net change		326,578	234,695	235,760	934,046	934,046

Notes:
1. The positive effects are induced by the government investments in forest protection and management, whereas the negative effects are induced by the logging bans and harvest reductions. The positive effects are underestimated due to the lack of consumption data for loggers, who are compensated by the NFPP. After 2003, the negative effects induced by the logging bans and harvest reductions will be zero.
2. See discussion on pp. 4–5 for logging bans and harvest restrictions over time; and see Table 14.2 for investments in forest protection, management, and expansion.
3. The results are reported in annual changes for comparing with the base case.

Table 14.8 Annual Changes of Wages and Salaries in the Forest Sectors Induced by the NFPP (Unit: 1000 yuan)

Sector		1998	1999	2000	2005	2010
Forest management	Increase	1,507,780	2,803,400	3,499,279	3,686,094	3,686,094
	Decrease	100,938	238,408	326,852	0	0
	Net change	1,406,842	2,564,992	3,172,427	3,686,094	3,686,094
Logging and hauling	Increase	1,957	3,356	4,189	4,412	4,412
	Decrease	253,868	983,267	1,348,037	0	0
	Net change	−251,911	−979,911	−1,343,848	4,412	4,412
Sawmilling and panel production	Increase	1,043	1,665	2,078	2,189	2,189
	Decrease	554	1,308	1,793	0	0
	Net change	489	357	285	2,189	2,189
Production of furniture and woodworks	Increase	647	1,059	1,322	1,392	1,392
	Decrease	314	743	1,018	0	0
	Net change	333	316	304	1,392	1,392
Paper and paperboard production	Increase	3,791	6,242	7,791	8,207	8,207
	Decrease	1,733	4,093	5,612	0	0
	Net change	2,058	2,149	2,179	8,207	8,207
Subtotal net change		1,157,811	1,587,903	1,831,347	3,702,294	3,702,294
Other sectors	Increase	284,282	457,279	570,787	601,260	601,260
	Decrease	133,498	315,312	432,286	0	0
	Net change	150,784	141,967	138,501	601,260	601,260
Total net change		1,308,595	1,729,870	1,969,848	4,303,554	4,303,554

Notes:
1. The positive effects are induced by the government investments in forest protection and management, whereas the negative effects are induced by the logging bans and harvest reductions.
2. See discussion on pp. 4–5 for logging bans and harvest restrictions over time; and see Table 14.2 for investments in forest protection, management, and expansion.
3. The results are reported in annual changes for comparing with the base case
4. All monetary values are in 1997 Chinese yuan.

Table 14.9 Changes of Value-Added in the Forest Sectors Induced by the NFPP (Unit: 1000 yuan)

Sector		1998	1999	2000	2005	2010
Forest management	Increase	1,707,467	3,174,676	3,962,716	4,174,272	4,174,272
	Decrease	114,306	269,982	370,139	0	0
	Net change	1,593,161	2,904,694	3,592,577	4,174,272	4,174,272
Logging and hauling	Increase	4,201	7,204	8,992	9,472	9,472
	Decrease	671,448	1,585,909	2,174,246	0	0
	Net change	−667,247	−1,578,705	−2,165,254	9,472	9,472
Sawmilling and Panel Production	Increase	2,289	3,652	4,559	4,803	4,803
	Decrease	1,215	2,869	3,933	0	0
	Net change	1,074	783	626	4,803	4,803
Furniture and solid wood products	Increase	1,293	2,117	2,642	2,783	2,783
	Decrease	629	1,485	2,036	0	0
	Net change	664	632	606	2,783	2,783
Paper and paperboard products	Increase	6,664	10,971	13,694	14,425	14,425
	Decrease	3,046	7,195	9,864	0	0
	Net change	3,618	3,776	3,830	14,425	14,425
Subtotal net change		931,270	1,331,180	1,432,385	4,205,755	4,205,755
Other sectors	Increase	555,697	893,630	1,115,452	1,175,003	1,175,003
	Decrease	273,013	644,836	884,056	0	0
	Net change	282,684	248,794	231,396	1,175,003	1,175,003
Total net change		1,213,954	1,579,974	1,663,781	5,380,758	5,380,758

Notes:
1. The positive effects are induced by the government investments in forest protection and management, whereas the negative effects are induced by the logging bans and harvest reductions.
2. See discussion on pp. 4–5 for logging bans and harvest restrictions over time; and see Table 14.2 for investments in forest protection, management, and expansion.
3. The results are reported in annual changes for comparing with the base case.
4. All monetary values are in 1997 yuan.

five forest sectors – forest management, logging and hauling, sawmilling and panel production, furniture and solid wood products, and paper and paperboard making – by 5.8 billion Yuan and the overall economy by 8.9 billion Yuan by 2010. Employment will increase by 0.84 million in the forest sectors and by 0.93 million in the whole economy.

Therefore, if properly implemented, the positive impacts of the NFPP would much more than offset the negative consequences of the logging bans. It is also clear that the potentially tremendous contributions to mitigating the problems of water runoff, soil erosion, flooding, and biodiversity loss are associated with the enormous expansion of forest protection and management. These environmental impacts should be evaluated as soon as possible.

In short, the NFPP can be a great environmental and economic policy. Further, it seems worthwhile to make a trade-off between the short-run revenues and jobs from exploiting the natural forests and the long-run sustainable development. However, the challenges are to truly transform the loggers into tree planters and forest managers, to ensure the financial and institutional commitments made by the central and regional governments to be materialized, and to complete the necessary structural adjustments in an efficient and coherent fashion.

Nevertheless, care should be taken in reaching any definite conclusions from our findings. In addition to the shortcomings of our data – the 1997 national I–O table and forestry statistics may not be as consistent or reliable as we wish, the I–O model used in this study is static and thus restrictive. It assumes fixed input substitution as well as exogenous determination of the final demand. Further, our estimation was based on the routine of aggregating the 124 sectors first and then creating the inverse matrix of the aggregated sectors and carrying out the necessary calculations. We did not pursue the other alternative – creating the inverse matrix of the 124 sectors and conducting the involved calculations first and then aggregating the results to a smaller number of sectors for presentation. Therefore, the assessed policy impacts may not be very accurate.

There are other approaches, such as the computable general equilibrium (CGE) and dynamic I–O (D/I–O) models, which can be used to mitigate some of the concerns or relax some of the assumptions. However, adopting those approaches is not cost free, too. For instance, while a D/I–O model can explicitly distinguish between investment and consumption demands and make either or both of them endogenously determined (Leontief, 1936; Schaffer, 1999), it requires more data and time to develop. Also, it may not be easy to forecast the future values of these variables. As to the CGE model, its advantage of capturing the general equilibrium (rather than only the partial equilibrium) effects must be weighed against its assumptions of the existence of such an equilibrium and fixed functional forms and price elasticities, as well as its added needs for data and analytic work (Kaimowitz & Angelson, 1998). It is based on these considerations that we decided to undertake a static I–O analysis in this study. On balance, we believe that our work has provided some interesting estimates of the NFPP impacts, which should significantly advance our understanding of the long-term policy-induced changes in the forest sectors and the economy.

Finally, it should be pointed out that our I–O model can be extended along other meaningful directions as well. For one, they can be expanded into an environmental I–O (EIO) analysis by augmenting the transaction matrix with additional rows and columns to represent the material flows or status changes for soil erosion and water runoff. Then, it will be practical to assess the environmental effects induced by the NFPP in a unified framework. Additionally, if regional I–O tables are available, an I–O analysis can be done at the regional level so that close attention can be given to those regions most affected by the NFPP. Fortunately, we have obtained China's first regional I–O table, which was compiled by the SSB in 2003. It will be exciting to enhance and expand our current assessment by taking its advantage.

Acknowledgments This article was originally published in *Environment and Development Economics* in 2006 [11(6): 769–788]. The authors appreciate comments made then by two anonymous reviewers and the Editors, as well as those by Gary Bull, Bill Hyde, Zhou Li, Can Liu, Gary Man, Xiufang Sun, Jennifer Turner, Andy White, Qing Xiang, Jintao Xu, and Lei Zhang. They are also grateful to Yaxiong Zhang from China's National Center of Economic Information, who provided the input–output tables. This project was partially supported by the US National Science Foundation, the Forest Trends, and USDA Forest Service Office of International Programs.

References

Chen, L., Xiang, C., Liu, X., & Mu, K. (2001). Case studies on the Natural Forest Protection Program in Sichuan province (working report). Beijing: CCICED Forest and Grassland Task Force.

China Forestry Statistics Yearbook. (1997–1999). Beijing: China Forestry Press.

China Statistics Bureau. (2000). *The 1997 National input–output table*. Beijing: China Statistics Press.

Davis, L. S., Johnson, K. N., Bettinger, P. S., and Howard, T. E. (2000). Forest Management. Boston, MA: McGraw Hill Higher Education.

Forest and Grassland Taskforce. (2003). *In pursuit of a sustainable green west* (Newsletter, January). Beijing.

Fullen, M. A., & Mitchell, D. J. (1994). Desertification and reclamation in north-central China. *Ambio, 23*(2), 131–135.

Hubacek, K., & Sun, L. X. (1999). *Land-use change in China: A scenario analysis based on input–output modeling*. Luxemburg, Austria: International Institute for Applied Systems Analysis.

Kaimowitz, D., & Angelson, A. (1998). *Economic models of tropical reforestation: A review*. Bogor, Indonesia: Center for International Forestry Research.

Leontief, W. W. (1936). Quantitative input and output relations in the economic systems of the United States. *The Review of Economic Statistics, 18*(3), 105–125.

Li, Z. (2001). *Conserving natural forests in China: historical perspective and strategic measures* (working report). Beijing: Chinese Academy of Social Sciences.

Liang, C. (1998). Flood investigations continue. *China Daily* (November 2).

Liu, C. (2002). *An economic and environmental evaluation of the Natural Forest Protection Program* (working paper). Beijing: SFA Center for Forest Economic Development and Research.

Loucks, C. J., Lü, Z., Dinerstein, E., Wang, H., Olson, D. M., Zhu, C. Q., et al. (2001). Giant pandas in a changing landscape. *Science, 294*: 1465.

Miller, R., & Blair, P. (1985). *Input–output analysis: Foundations and extensions*. Englewood Cliffs, NJ: Prentice-Hall, Inc.

Munn, I. A. (1998). *The impact of the forest products industry on the Mississippi economy: an input–output analysis*. (FWRC Research Bulletin #FO 087). Starkville, MS

Rawski, T. G. (2001). China's GDP statistics: a case of caveat lector? *China Economic Quarterly,* *5*(1), 18–22.

Schaffer, A. W. (1999). *Regional impact models.* Working report, Morgantown, WV: Regional Research Institute.

Shen, M. (2000). *How the logging bans affect community forest management: Aba prefecture, North Sichun.* Working paper, Chengdu: Institute of Rural Economy, Sichuan Academy of Social Sciences.

Sheng, M. Y. (2002). *A study of the Natural Forest Protection Program and the Slope Land Conversion Program.* Chengdu: Social Science Academy of Sichuan Province.

State Forestry Administration (SFA). (2000–2002). *China forestry development report.* Beijing: China Forestry Press.

State Statistics Bureau. (1998). The 1997 National Input-Output Table. Beijing: China Statistics Press.

Studley, J. (1999). Forests and environmental degradation in Southwest China. *International Forestry Review, 1*(4), 260–265.

World Bank (1994). *China: Forest Resource Development and Protection Project.* Washington, DC:

Wu, W. (2002). *Economic impact of the Texas forest sector.* (Forest Resource Development Publication No. 161). College Station, Texas: Texas Forest Service.

Xu, J. T., Yin, R. S., Li, Z., & Liu, C. (2006). China's ecological rehabilitation: Progress and Challenges. *Ecological Economics, 57* (4): 595–607.

Xu M., Qi, Y., & Gong, P. (2000). China's new forest policy. *Science, 289,* 2049–2050.

Yu, Y., Xie, C., Li, C., & Chen, B. (2002). *The NFPP and Its Impact on Collective Forests and Community Development.* Report commissioned by the Forest and Grassland Taskforce, Beijing.

Zhang, Y., Buongiorno, J., & Zhang, D. L. (1997.) China's economic and demographic growth, forest products consumption, and wood requirements: 1990 to 2010. *Forest Products Journal, 47*(4), 27–35.

Zhang, P. C., Shao, G. F., Zhao, G., Le Master, D. C., Parker, G. R. Dunning J. B. Jr., et al. (2000). China's forest policy for the 21st century. *Science, 288,* 2135–2136.

Zhao, G., & Shao, G. F. (2002). Logging restrictions in China: a turning point for forest sustainability. *Journal of Forestry, 100*(4), 34–37.

Index

A
Adaptive capability, 16
Administrative budgeting and capacity, 16
Aerial seeding, 4, 6, 10, 11, 237
Afforestation, 3, 6, 8, 10, 11, 13, 14, 31, 33,
46, 80, 82, 95, 104, 108, 110, 114, 177,
203, 237, 239
Age structure, 3, 25
Animal husbandry, 9, 72, 90, 93, 159, 160,
161, 163, 166, 167, 168, 169, 171, 172,
204, 226, 231
Annual discharge, 116, 118
Annual erosion, 116, 122, 123, 126, 128
Annual precipitation/rainfall, 33, 57, 101, 105,
109, 114, 117, 119, 121, 123, 124
Annual runoff, 127
Annual temperature, 57, 104, 105, 106, 109,
114
Artificial planting, 4, 6, 10, 236
Atmosphere, 99, 100, 107, 109
Attrition rate, 137

B
Beijing, 2, 4, 6, 9, 12, 22, 27, 32, 203, 204,
208, 220, 238
Biodiversity hotspot, 70, 101
Biogeochemical and biophysical processes,
100
Biological diversity, 41, 45, 46
Biophysical and socioeconomic factors, 75, 86,
206
Biosphere, 99, 100, 109

C
Canopy height, 126
Canopy interception, 126, 127
Capacity building, 14
Carbon budget, 100
Carbon cycle, 107

Carbon sequestration, 22, 23, 107, 109, 131
Carbon sink and source pattern, 107–109
Carbon stock, 44, 45, 101, 106–107, 108, 109
Carbon storage and flux, 46
Cash crops, 121, 201, 217
Causal relationships, 87
China, 1–18, 21–36, 39–52, 55–66, 70, 71, 75,
82, 86, 87, 89, 100, 101, 102, 103, 113,
114, 115, 116, 124, 131–155, 159–172,
177, 178, 180, 196, 201–217, 220, 221,
230, 235–253
China Forestry Development Report, 237, 239,
244
China Remote Sensing Satellite Ground
Station, 58
Chi-square test, 93
Classification accuracy, 60, 120
Climate change, 46, 51, 86, 97, 109
Coefficient of efficiency, 117, 118, 126
Collective forests, 15, 31, 76, 80, 82, 177, 179,
180, 195, 196, 197
Collectivization, 3
Commercial forests, 5, 17, 29
Community participation, 14, 15, 26, 27, 49
Comparative advantage, 80, 171
Competition control, 14
Confusion matrix, 60
Conservation payment, 131–155
Conservation Reserve Program (CRP), 26,
133, 153, 220
Conservation set-aside, 49, 132, 137
Contour ditch, 125
Control variables, 47, 142, 163, 206, 212
Cost effectiveness, 26, 28, 35, 41, 47,
49–50, 205
Counterfactual, 27, 144, 181, 184, 224
Coupled human and natural processes, 86
Cover factor, 117, 128, 129
Credit market, 134, 137

R. Yin (ed.), *An Integrated Assessment of China's Ecological Restoration Programs*,
DOI 10.1007/978-90-481-2655-2_BM2, © Springer Science+Business Media B.V. 2009